Plant Pathology in the 21st Century

Volume 11

Series Editor

Maria Lodovica Gullino, AGROINNOVA, Università degli Studi di Torino, Grugliasco, Turin, Italy

The aim of this Series is to highlight the latest international findings and advances in plant pathology and plant disease management, and plant pathology topic specialist, Congress and Workshop organisers, coordinators of broad International projects are invited to consult with the Series Editor regarding their topic's potential inclusion in the series.

Under an agreement with Springer, a Book Series based on the invited lectures at the 9th International Congress of Plant Pathology ICPP 2008, *Plant Pathology in the 21*st *Century* was initiated and four books covering key themes were published. Following on the procedure for the 2008 Congress, three additional volumes in the series were published on themes which were key topics at ICPP 2013, held in Beijing, China, while two new books are almost ready from ICPP 2018 held in Boston (SA). Moreover, more books were published in between Congresses, covering up-to-date topics in plant pathology. In light of the initial seven volumes' success, the ISPP has now concluded an agreement with Springer to broaden the scope of the Series and publish additional volumes.

The goal of the International Society for Plant Pathology (ISPP; www.isppweb.org) is to promote the global advancement of plant pathology and the dissemination of essential information on plant diseases and plant health management. This book Series looks of particular interest due to the upcoming International Year of Plant Health (2020).

More information about this series at http://www.springer.com/series/8169

Davide Spadaro • Samir Droby
Maria Lodovica Gullino
Editors

Postharvest Pathology

Next Generation Solutions to Reducing Losses and Enhancing Safety

Springer

Editors
Davide Spadaro
AGROINNOVA and DISAFA, Università degli Studi di Torino
Grugliasco, Turin, Italy

Samir Droby
The Volcani Center, Agricultural Research Organization
Rishon LeZion, Israel

Maria Lodovica Gullino
AGROINNOVA, Università degli Studi di Torino
Grugliasco, Turin, Italy

ISSN 2512-160X ISSN 2512-1626 (electronic)
Plant Pathology in the 21st Century
ISBN 978-3-030-56529-9 ISBN 978-3-030-56530-5 (eBook)
https://doi.org/10.1007/978-3-030-56530-5

This Springer imprint is published by the registered company Springer Nature Switzerland AG
The registered company address is: Gewerbestrasse 11, 6330 Cham, Switzerland

Preface

The 11th edition of the International Congress of Plant Pathology (ICPP-11) was held in Boston, USA, in 2018. Postharvest pathology is at the crossroad between food security and food safety, and it represents an essential element of plant pathology. This book consists of a collection of papers of the presentations given during two concurrent sessions dedicated to "*Pathogenicity and Resistance in Post-Harvest Diseases*" and "*Novel and Integrated Approaches to Control Post-Harvest* Diseases" and the workshop "*Current Issues in Food Safety and Post-Harvest Pathology*."

Losses caused by pests and diseases on fresh fruits, nuts, and grains along the supply chain (at harvest, storage, transit, and subsequent commercialization steps, before reaching the consumer) are not easily assessed. It is estimated to reach 25% of the total production in countries where postharvest facilities as well as handling and storage facilities are well developed. In developing countries, losses are often higher, exceeding 50%, because of the lack of adequate storage facilities and appropriate postharvest handling chains.

Microbial decay, caused by fungi and bacteria, is one of the main factors that determine losses compromising both the quality and safety of fruits, vegetables, nuts and grains. The development of an integrated approach for decay management, preharvest, harvest, and postharvest practices should be considered as essential components that influence the complex interaction between host, pathogen, and environmental conditions. Among postharvest practices, fruit treatments with fungicides are the most effective means to reduce decay. However, the wide consumption of high-quality fresh fruits and vegetables and the increased concerns over the possible toxicity of fungicide residues have led to the development of new alternative approaches for disease control, such as physical means and use of biocontrol agents and natural compounds. The implementation of these alternatives techniques often requires modifying the currently used postharvest practices and the development of new technologies or formulations for their applications.

Four hot topics, which have been identified in the area of postharvest diseases management, also represent the four parts of this book: i) Elucidation of the Complex Fruit-Pathogen Interactions, ii) Study of the Role of the Fruit Microbiome on the Development of Postharvest Diseases, iii) Interaction Between Postharvest

Losses and Production of Mycotoxins by Postharvest Pathogens, and iv) Development of Sustainable Strategies for the control of Postharvest Diseases.

Three chapters are dedicated to the investigation of fruit–pathogen interactions. In Chap. 1, Levin et al. present an extensive analysis of the potential effectors of *P.expansum* that may be involved in its pathogenicity on apple fruit and examine their role in suppressing fruit resistance mechanisms in the initial stages of infection. Torres et al. (Chap. 2) provide insights into the fruit–pathogen interactions, specifically in orange and apple fruit defense responses against compatible and nonhost pathogens, such as *Penicillium digitatum* and *P. expansum*. Plants possess an innate immune system with disease resistance genes encoding for proteins that recognize pathogen effectors during infection. Parada-Rojas and Quesada-Ocampo (Chap. 3) review the state of the art of an important class of resistance genes—nucleotide-binding and leucine-rich repeat domains (NLRs)—in sweet potato as well as novel methods to predict NLRs in plant genomes.

Two chapters deal with the role of the fruit microbiome on the development of postharvest diseases.

Accumulating evidence indicates that the composition of the microbiota inhabiting an organism (both endo- and epiphytically) can have a profound effect on host physiology and defense responses. The global effort to characterize the endophytic and epiphytic microbiome of apple fruit is described by Wisniewski et al. (Chap. 4). The host and microbiome have co-evolved to some extent, as suggested in the holobiont concept. The implications of greater knowledge of the apple microbiome that would facilitate better disease and cultural management strategies, cultivar breeding, and abiotic stress resistance are discussed. The distribution pattern, as well as the potential utilization of endophytic microbiota that are associated with internal tissues for managing pre and postharvest pathogens, is presented in Chap. 5 by Kumar et al.

Geisen and Schmidt-Heydt (Chap. 6) develop the biological role of mycotoxin synthesis in fungi. Mycotoxins are described as pathogenicity factors of plants, colonization factors in the postharvest environment, or as specific adaptation factors for challenging and stressful environments. The overview of the interplay of the postharvest environment and mycotoxin biosynthesis opens the door to the other two chapters of the part Mycotoxins in Postharvest. Several species of *Aspergillus* and *Penicillium* contaminate nuts during the supply chain. It is, therefore, necessary to develop harvesting methods, storage conditions, processing techniques, and detoxification protocols for the management of fungal growth and mycotoxin contamination. The efficacy, the benefits, and the drawbacks of drying technologies, roasting, and cold plasma in reducing mycotoxin contamination are described by Spadaro et al. (Chap. 7). Kumar Solanki et al. (Chap. 8) deal with the role of the wheat microbiome in relation to the mycotoxin occurrence in stored grain. A better knowledge of the composition and dynamics of wheat-grain-associated microbiota is needed to identify novel beneficial microorganisms that may improve crop health and suppress the growth of potential pathogens in a sustainable manner.

The fourth part, dedicated to the development of sustainable strategies for postharvest disease control, is composed of four chapters. Adaskaveg et al. (Chap. 9) offer an overview of the chemical management of postharvest diseases of subtropical and tropical fruits, such as citrus, pomegranates, pineapples, and avocado. For subtropical crops, new fungicides have been introduced to manage major diseases, ranging from conventional (propiconazole), reduced-risk (pyrimethanil, azoxystrobin, fludioxonil) to bio-pesticide (natamycin), and also pre-mixtures. Postharvest fungicide applications with aqueous recycling drenches are one of the most effective strategies to prevent fruit decay. Adaskaveg et al. (Chap. 10) describe different sanitation practices to recycle fungicide drenches. Among alternative means to chemical fungicides to control postharvest decay of fresh horticultural products, low-toxicity chemicals, classified as food additives or Generally Recognized As Safe (GRAS) compounds, are of interest because of their low toxicological effects on mammals and minimal impact on the environment. Palou and Pérez-Gago (Chap. 11) introduce the reader to GRAS salts as a viable alternative for postharvest preservation of fresh horticultural products. Finally, Ippolito et al. (Chap. 12) focus on the advantages and drawbacks of electrolyzed water as a novel broad-spectrum sanitizer for postharvest disinfection to maintain commodities and facilities free from fungal and bacterial pathogens.

We hope this book will be helpful to researchers, extension services, and technicians working in the postharvest industry as well as students interested to this field.

Fruit–pathogen interactions, fruit microbiome, mycotoxins, and postharvest pest and disease management strategies are the keywords for this forefront book that provides interesting hints and suggestion to further develop the challenging discipline of postharvest pathology. Next Generation Solutions for reducing postharvest losses and enhancing food safety are at your fingertips!

Grugliasco, Turin, Italy Davide Spadaro
Rishon LeZion, Israel Samir Droby
Grugliasco, Turin, Italy Maria Lodovica Gullino

Contents

Chapter 1
Role of Effector Proteins in the Virulence of *Penicillium expansum* on Apple Fruit

Elena Levin, Samir Droby, Michael Wisniewski, and Christopher Dardick

Abstract *Penicillium expansum* is a ubiquitous postharvest pathogen that infects a wide range of fruits, including pome and a variety of stone fruit. Virulence factors that mediate pathogenicity in postharvest-decay fungi have not been fully defined, especially in regard to if and how the pathogen is able to modulate fruit defense mechanisms, particularly at the initial stages of infection. In this chapter, we present an extensive analysis of the potential effectors of *P.expansum* that may be involved in its pathogenicity on apple fruit and examine their role in suppressing fruit resistance mechanisms in the initial stages of infection. The ability of *P. expansum* to secrete factors that are able to down-regulate ROS production was initially assessed. Results demonstrated that there are proteins secreted by *P. expansum* that are able to down-regulate host-response to presence of the pathogen. A combination of different bioinformatic and genetic approaches were used to identify and characterize potential effectors in *P. expansum*. In particular, the small cysteine-rich proteins, *Pescr1* and *Pescr2*, were investigated. Additionally, the role of LysM proteins, NLPs, and one proteolytic enzyme, peptidase S8, in the virulence of *P. expansum* was also examined.

Keywords NLP effectors · LysM effectors · Pathogenicity · Necrotrophic fungi · Small · Cysteine-rich proteins · PePRT protein

E. Levin · S. Droby (✉)
Department of Postharvest Science, ARO, The Volcani Center, Rishon LeZion, Israel
e-mail: elenal@volcani.agri.ov.il; samird@volcani.agri.gov.il

M. Wisniewski · C. Dardick
Appalachian Fruit Research Station, U.S. Department of Agriculture – Agricultural Research Service, Kearneysville, WV, USA
e-mail: Michael.Wisniewski@ars.usda.gov; Chris.Dardick@ars.usda.gov

D. Spadaro et al. (eds.), *Postharvest Pathology*, Plant Pathology in the 21st Century 11, https://doi.org/10.1007/978-3-030-56530-5_1

Introduction

Penicillium is a fungal genus containing 354 recognized species (Visagie et al. 2014) that are distributed globally, where they inhabit a large variety of substrates, from soil to foods. *Penicillium expansum* is a postharvest, necrotrophic pathogen that infects a wide range of fruits, including pome and a variety of stone fruit (e.g. peaches, nectarines, plums, and apricots) (Cappellini et al. 1987; Jurick et al. 2011). *P. expansum* is regarded as the most important postharvest pathogen of apple fruit and is of great concern to the fruit processing industry due to its ability to synthesize toxic metabolites such as patulin (Wouters and Speijers 1996; McCallum et al. 2002). Direct fruit losses due to the development of decay during long-term storage of apple can be considerable and dependent on the cultivar, postharvest handling processes and storage conditions. Infection of apples by *P. expansum* usually takes place through wounds inflicted during fruit harvesting and subsequent postharvest handling. Spores germinate in the wounds and colonize the surrounding tissue within 24 h. A visible breakdown of the tissue surrounding the infection site occurs within 48 h. A unique feature of *P. expansum* compared to other postharvest pathogens is its ability to grow at low temperatures. This ability allows it to survive and grow during low–temperature storage of fruit. *P. expansum,* commonly referred to as blue mold, is a significant problem in production systems that do not use pre- or postharvest fungicides (Tahir et al. 2015).While apple producers rely on the use of synthetic fungicides for decay control, future use of postharvest fungicides will be restricted or banned by many countries (Droby et al. 2009). Alternative postharvest strategies, based on biological approaches, are being developed but are currently very limited due to their reduced efficacy under commercial conditions. Although one of the best means to control plant diseases is the use of resistant cultivars, little attention has been devoted to postharvest disease resistance in apple breeding programs (Janick et al. 1996). This is due both to a lack of sources of genetic resistance in domesticated apples and the current prohibitive cost of maintaining trees in trial plots for several years. Hence, the identification of resistance as a heritable trait would represent a significant accomplishment. To this purpose, the identification of DNA markers for resistance to post-harvest decay would increase the feasibility of breeding resistant cultivars through marker-assisted selection (MAS) of seedlings prior to field planting. In this regard, Norelli et al. (2017) tested the hypothesis that the observed resistance of wild *Malus sieversii* accessions PI613981 to infection by *P. expansum* is due to the effect of specific genetic loci. The analysis of a *M.* × *domestica* 'Royal Gala' X *M. sieversii* PI613981 mapping population (GMAL4593) identified two QTLs located on linkage groups (LG) 3 and LG10 that were associated with resistance to *P. expansum* and were designated qM-Pe3.1 and qM-Pe10.1, respectively. A DNA test for qM-Pe3.1, which will greatly improve the efficiency of breeding apple for blue mold resistance, has been developed and is currently being used to identify and introgress predicted blue mold resistance in progeny segregating for qM-Pe3.1 (Norelli et al. 2017). A better understanding of the mechanism of pathogenicity of *P. expansum* on apple and how the pathogen is able to overcome innate immunity on fruit would provide new insights that could be used for developing innovative management strategies to control this postharvest pathogen.

Plant Defense Mechanisms

Plants respond to pathogen attacks using different defense mechanisms. Innate or preformed defense mechanisms involve structural barriers such as a waxy cuticle and the cell wall or preformed antimicrobial compounds that function to prevent the colonization of the tissue (Jackson and Taylor 1996; Osbourn 1996). Once the structural barriers of a host have been breached, plants respond to infection using a two-branched innate immunity system (Jones and Dangal 2006). The first branch recognizes and responds to pathogen-associated-molecular-patterns (PAMPs), molecules that are common to many classes of microbes, including non-pathogens. This type of immune response is referred to as PAMP-triggered immunity (Doehlemann et al. 2009; Nürnberger and Kemmerling 2009). Pathogens, however, have evolved the use of effectors that interfere with PAMP-triggered immunity and results in effector-triggered susceptibility (Boller and He 2009). In the second branch of innate immunity systems, a given pathogen effector is recognized by a resistance protein resulting in effector-triggered immunity (ETI). This second level of immunity can itself be suppressed by other pathogen effectors (Houterman et al. 2008), resulting in the evolution of complex host-microbe interactions (Chisholm et al. 2006).

Pathogenicity of *P. expansum* on Fruit

Fungal-plant pathogens have developed several mechanisms of pathogenicity that enhance their virulence (Perez-Nadales et al. 2014). In regard to the *Penicillium* spp. host-pathogen interaction, wounding of the fruit surface, secreted nutrients, and volatiles stimulate spore germination (Droby et al. 2008). Spore germination is followed by penetration and colonization of the fruit tissue. *P. expansum* is able to cause decay at the low temperatures used for fruit storage, and over-mature or long-stored fruits are especially susceptible to infection. In most cases, necrotrophic fungal plant pathogens promote host cell death by secreting toxic extracellular proteins and secondary metabolites and then utilize the nutrients released from dead host cells for growth (Van Kan 2006). In the early stages of infection and disease development, a wide range of virulence mechanisms are employed by these pathogens to overcome the innate and induced resistance responses of their hosts. These include the secretion of diverse phytotoxic compounds, cell-degrading enzymes, and proteinaceous effectors that function in manipulating host defense mechanisms (Wang et al. 2014). Only a relatively few protein effectors have been identified in necrotrophic pathogens, however, and the molecular interactions between these effectors and host plants, in most cases, remain to be elucidated.

While the availability of the genome sequences of several of *P. expansum* strains can be used in the identification of virulence genes (Ballester et al. 2015), there is general lack of knowledge about virulence elements that determine pathogen aggressiveness in *P. expansum*, including information on effectors that can subvert fruit defenses.

Several reports on the potential role of different factors that mediate pathogenicity and virulence in *P. expansum* on apple fruit have been published in recent years. Acidification of host tissue leading to enhanced production of plant cell-wall-degrading fungal enzymes, as well as the production of patulin are among the suggested factors influencing the pathogenicity and virulence of *P. expansum* (Prusky et al. 2004; Sanzani et al. 2012; Barad et al. 2014; Vilanova et al. 2014). Patulin is a secondary metabolite and its role in the development of blue mold decay on apples remains equivocal. While some authors have implicated its involvement in pathogenicity (Sanzani et al. 2012; Barad et al. 2014), other authors have reported the lack of an association between virulence and patulin production in *P. expansum* (Ballester et al. 2015; Li et al. 2015). It appears that, although patulin could not be required to infect apples, it may act as an apple cultivar-dependent aggressiveness factor (Snini et al. 2016).

Previous studies have linked the secretion of polygalacturonases, glutamate decarboxylases, calmodulin or C-4 methylesterol oxidases with the pathogenicity of *P. expansum* in apples by characterizing genes that are differentially expressed during the infection process (Sánchez-Torres and González-Candelas 2003). Qin et al. (2007) also linked polygalacturonases to the pathogenicity of *P. expansum* by analyzing its cellular and extracellular proteomes. Moreover, Tian et al. (2013) reported that proteins involved in antioxidant metabolism, such as catalases and superoxide dismutases, are related to the pathogenicity of blue mold. The hypothesis that *P. expansum* has the ability to acidify the environment in colonized tissue by the accumulation of D-gluconic acid has also been proposed as a mechanism of *P. expansum* pathogenicity in apples (Vilanova et al. 2014).

Fruit pH has been suggested to be an important factor in postharvest diseases due to its direct effect on conidial germination (Pelser and Eckert 1977) and on the regulation of virulence factors (Alkan et al. 2013). These authors reported that when *P. expansum* colonizes apple fruit tissues, it acidifies its local surroundings causing the enhancement of its pathogenicity by activating pathogenicity factors (Alkan et al. 2013).

Virulence factors that mediate pathogenicity in postharvest-decay fungi have not been fully defined, especially in regard to if and how the pathogen is able to modulate fruit defense mechanisms, particularly at the initial stages of infection. While pathogen effectors responsible for suppressing host defense mechanisms are believed to play a major role in the pathogenicity of necrotrophic pathogens, their potential involvement in the apple-*P. expansum* system has not been fully elucidated.

Fungal Effectors and Their Role in the Pathogenicity of Necrotrophic Fungi

Fungal plant pathogens are classified into three groups based on their mechanisms of pathogenicity. The groups are: biotrophs (obtaining nutrients from the host directly, but not killing the tissue); hemi-biotrophs (fungi that have an early biotrophic stage but then become necrotrophic); and necrotrophs (fungi that kill cells and tissues to obtain nutrients). Most of these fungal pathogens utilize effectors to suppress innate and induced resistance mechanisms, which allows them to invade host tissues (Kleemann et al. 2012).

Pathogen effectors are defined as molecules that manipulate host cell structure and have dual function in a manner that in cases it facilitates infection and in others triggers defense responses (Kamoun 2006). Necrotrophic fungi have been generally considered pathogens that rely on a plethora of non-host-specific mechanisms to infect their host. Studies on effectors in necrotrophic fungi have demonstrated, however, that they are capable of selectively suppressing host defenses before an effective defense response can be initiated by the plant host. However, the mode of action of only few of these effectors have been studied. PtrToxA, a proteinaceous effector in *Pyrenophora tritici-repentis* the causal agent of tan spot disease of wheat (*Triticum aestivum L.*), was demonstrated to be present within living mesophyll cells of sensitive wheat cultivars. PtrToxA is targeted to the chloroplast and apparently disrupts photosynthetic electron transport, leading to ROS accumulation and plant cell death upon light exposure (Tan et al. 2010). Weiberg et al. (2013) demonstrated that *Botrytis cinerea,* a highly destructive plant necrotrophic pathogen, is able to suppress host immunity in *Arabidopsis* and tomato plants through small RNAs that highjack RNA interference (RNAi) machinery and selectively silence host immunity genes. In other studies, host-selective toxins were identified that function in effector-triggered susceptibility (Friesen et al. 2007, 2008; Manning et al. 2007; Abeysekara et al. 2009; Liu et al. 2009, 2012). LysM effectors have been shown to contribute to pathogenicity and virulence in several pathogen-plant systems (Bolton et al. 2008; De Jonge et al. 2010). Necrosis and ethylene-inducing peptide 1 (NEP1)-like (NLP) effectors, found in many plant pathogens, have also been shown to play a role in the pathogenicity of necrotrophic fungi (Schouten et al. 2008; Dallal Bashi et al. 2010). Other effectors, such as the PeSte12 transcription factor, were reported to affect lesion development of *P. expansum* on apple (Sánchez-Torres et al. 2018). During axenic growth, PeSte12 appears to act as a negative regulator of genes related to detoxification, ATPase activity, protein folding, and basic metabolism. Knocking out *PeSte12* gene significantly affected the rate of decay development on apple fruit stored at 0 °C. No significant effect, however, was observed when apple fruits were inoculated with knockout mutants and stored at 20 °C (Sánchez-Torres et al. 2018).

P. expansum Secretes Factors That Down-Regulate ROS Production in Apple During Infection

An extensive analysis of potential effectors of *P.expansum* was conducted to identify potential effectors involved in the pathogenicity of *P.expansum* on apple fruit and to examine their role in suppressing fruit resistance mechanisms during the initial stages of infection. The first stage of this study was directed at understanding the response of fruit tissue to elicitors secreted by *P. expansum* mycelia. Fragments of fungal cell walls, such as chitin oligomers, are known to act as elicitors that trigger ROS production in host tissues. The induced ROS usually serves as a signal to trigger downstream pathways leading to the activation of resistance mechanisms (Lamb and Dixon 1997; Torres et al. 2006; Torres 2010). Necrotrophic pathogens usually secrete effectors that suppress ROS production during the initial stages of the infection process to inhibit the activation of induced resistance responses by the host and thus allow it to establish infection (Wang et al. 2014). In our study, apple wounds were inoculated with 50 mg/ml of a preparation of *P. expansum* cell walls mixed with culture filtrate that was derived from the fungus growing in an apple juice medium. This was done to determine if *P. expansum* secretes factors that are able to down-regulate ROS production. ROS production was measured in wounded tissues using fluorescence staining. As illustrated in Fig. 1.1, wounds treated with the fungal cell wall preparation plus the addition of culture filtrate markedly reduced ROS production in host tissues, relative to treatments with fungal cell walls alone. This suggests that materials secreted into the culture filtrate play an active role in down-regulating ROS production in apple tissue. Subsequently, culture filtrate was first boiled and then mixed with cell walls before being administered into apple wounds to determine if the ROS-suppressive factors were proteins. In this case, results indicated that the fluorescence level in apple tissues was significantly higher relative to the fluorescence in wounds inoculated with cell walls with not boiled culture filtrate (Fig.1.1). The results of this experiment suggest that proteins secreted by *P. expansum* are able to inhibit or prevent host-response to fungal infection. This effect was also described by Buron-Moles et al. (2015) in apples inoculated with host and non-host species of *Penicillium.* In that study, production of hydrogen peroxide (H_2O_2) in immature apple fruit tissues that was induced by wounding was suppressed by *P. expansum* to much greater extent than it was by the non-host pathogen, *P. digitatum* (Buron-Moles et al. 2015).

Discovery and Functional Analysis of Effectors in Plant Pathogenic Fungi

A combination of biochemical, genetic, and bioinformatic strategies have often been used for the discovery and characterization of effector genes in plant pathogenic fungi (Kamoun 2006; Raffaele et al. 2010). Recent advances in genomics,

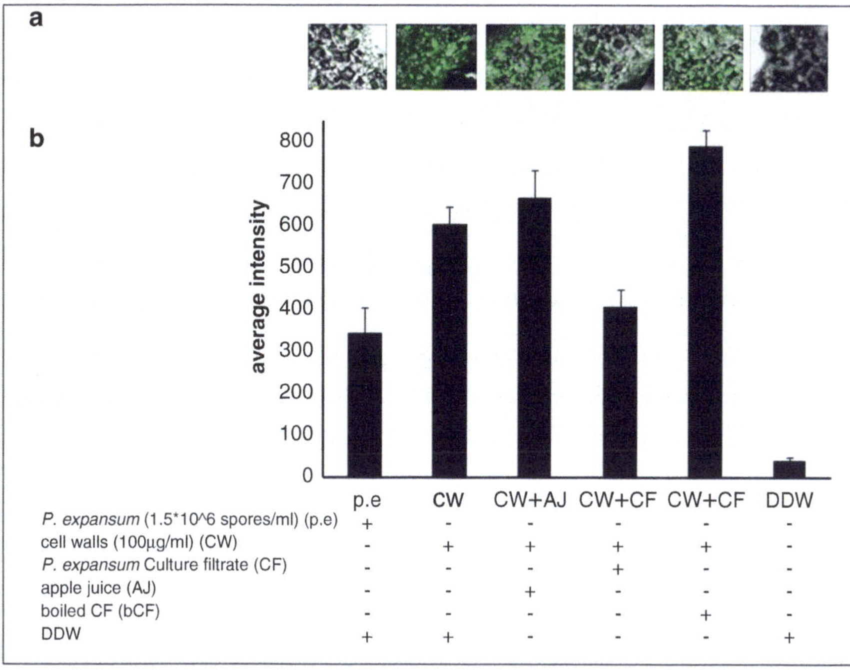

Fig. 1.1 Effect of the factors secreted by *P. expansum* on the ROS production in apple wounds. (**a**) confocal microscope image of ROS production in apple wounds inoculated with *P. expansum* spores (p.e), *P. expansum* cell walls fragments (CW) with or without addition of *P. expansum* culture filtrate (CF) visualized by DCF staining, (**b**) quantification of the fluorescence intensity in the images

proteomics, and bioinformatics have facilitated the implementation of functional genomic pipelines to identify effectors from nucleotide and protein sequence data sets. As suggested by Kamoun (2006), a typical pipeline consists of two major steps: use of data mining tools to identify candidate genes that fulfill a list of specific criteria, followed by analysis and validation of these candidate genes by functional assays, such as expression *in planta* and evaluation of effector-like activity. This approach was successfully used to develop a pipeline for the discovery of *Phytophthora infestans* effectors (Torto et al. 2003; Raffaele et al. 2010). Identification of candidate secreted proteins was facilitated by the fact that in Oomycetes, as in other eukaryotes, most secreted proteins are exported through a common secretory pathway that utilizes short, N-terminal, amino-acid sequences known as signal peptides (Torto et al. 2003). Signal peptides are highly degenerate and cannot be identified using DNA hybridization or PCR-based techniques. Nonetheless, computational tools, particularly SignalP software (Nielsen 2017), can be used to assign a signal peptide annotation to unknown amino acid sequences (Menne et al. 2000; Schneider and Fechner 2004).

One of the problematic challenges facing researchers in the identification of potential effector proteins is the lack of common features in effector genes/proteins within and across species. Since there is an ever-increasing number of sequenced genomes of plant pathogenic fungi, computational tools have become a widely used approach for effector-prediction (Sperschneider et al. 2013, 2014, 2015; Syme et al. 2013). Criteria such as presence of a secretion signal, species-specificity, small size, a large number of cysteine residues, presence of specific protein domains and motifs found in known fungal effectors, signatures of positive selection, recent gene duplication, and homology to proteins experimentally proven to affect the outcome of pathogen-host interactions have all been used to identify potential effectors (Guyon et al. 2014; Levin et al. 2019a).

In the present study, nine potential effectors of *P. expansum* were examine, four of which were detected by a pipeline developed by Levin et al. (2019a). The remaining five effector candidates were members of known effector families reported to contribute to the virulence of other plant-pathogenic fungi.

Small, Cysteine-Rich Proteins in *P. expansum*

Small cysteine-rich secreted proteins with unknown function are among the most established parameters for the identification of potential effectors (Sperschneider et al. 2015). Analysis of the *P. expansum* secretome revealed 25 small (< 300 aa) cysteine-rich (>4) proteins with unknown function. Three of them exhibited a high level of expression during apple infection and one was induced during spore germination (Levin et al. 2019a). The effect of two proteins on the virulence of *P. expansum* was examined: *Pescr1* (PEX2_056460), expressed in ungerminated spores, and *Pescr2* (PEX2_002470) expressed during apple infection.

Deletion of *Pescr1* had no effect on the radial growth rate and colony morphology of mutant strains of *P. expansum* when grown on PDA medium (Fig. 1.2a, b). The *Pecsr2* null mutant, however, exhibited a slightly lower rate of radial growth compared to the wild-type (WT) strain and the ectopic mutant. 'Golden Delicious' apples were inoculated with the two null mutants to examine the effect of the deletions on pathogenicity. The rate of lesion development and the percent of infected wounds were compared between null mutants, ectopic mutants, and the WT strain. No significant differences in infection incidence were observed between either of the mutants. As illustrated in Fig. 1.2c, however, lesion diameter caused by the *Pescr2* mutant was significantly smaller than the lesion diameter induced by the WT strain and ectopic mutants. This result is similar to the results obtained for the disruption of *Ecp1* or *Ecp2* genes coding for cysteine-rich extracellular proteins that were abundantly secreted by *Cladosporium fulvum* during colonization of tomato leaves, which compromised its virulence (Laugé et al. 1997). Moreover, functional analyses of *F. oxysporum f. sp. lycopersici* cysteine-rich, secreted proteins encoded by avirulence (AVR) genes (*Avr1*, *Avr2*, and *Avr3*) revealed that Avr2 and Avr3, in addition to triggering resistance in tomato plants carrying the R genes *I-2* and *I-3*,

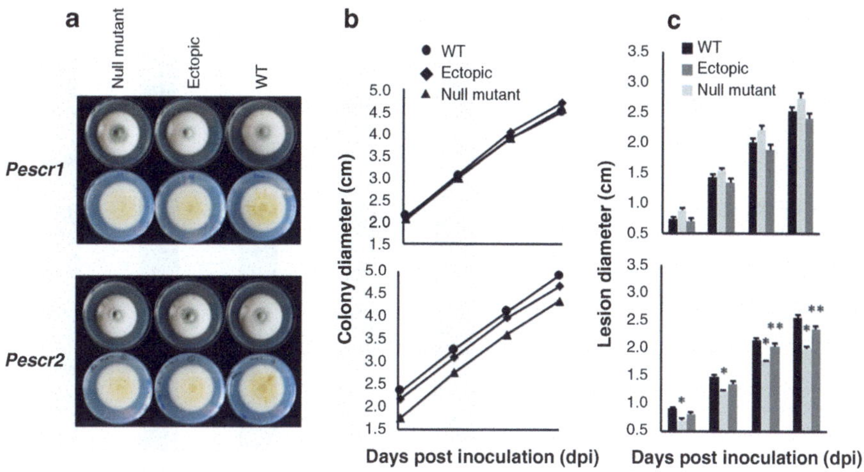

Fig. 1.2 Effect of the targeted deletion of genes coding for small cystein-rich proteins PeSCR1 and PeSCR2 on the growth rate, colony morphology and virulence. (**a**) colony morphology at 7 dpi, (**b**) radial growth on PDA medium, (**c**) lesion diameter during apple infection of the wild type (WT), extopic mutants (E), and null mutants. Radial growth and lesion diameter results are an average of three independent measurments

respectively, were both required for full pathogenicity insusceptible tomato lines (Rep et al. 2004; Rep 2005; Houterman et al. 2009). *F. oxysporum f. sp. lycopersici* effector Avr1, however, does not contribute to virulence insusceptible cultivars.

LysM Proteins in *P. expansum*

Genetic homologues of lysine motif (LysM) effectors have been identified in many fungal species, and were reported to have a major effect on virulence in several plant pathogens (Bolton et al. 2008; De Wit et al. 2009; De Jonge et al. 2010; Marshall et al. 2011; Mentlak et al. 2012; Takahara et al. 2016; Kombrink et al. 2017). Analysis of the genome of *P. expansum* revealed 18 genes possessing a LysM motif. Eleven of these LysM genes were predicted to also contain a secretion signal, while only four (*PeLysM1*, *PeLysM2*, *PeLysM3*, *PeLysM4*) were found to be actively transcribed during the early stages of the infection process of apple (Levin et al. 2017). A further analysis was conducted of the amino acid sequences of the LysM motif of these genes. An amino acid alignment of the LysM motifs in *P. expansum* with LysM domains from known fungal effectors revealed a high level of homology to the LysM domains of the *Trichoderma atroviride* TAL6 protein rather than other LysM effector proteins (Levin et al. 2017). Notably, TAL6 was demonstrated to be involved in self-signaling processes during fungal growth (Seidl-Seiboth et al. 2013).

The analysis of the knockout mutants of each of the four LysM genes included an examination of growth rates, colony morphology, sporulation, germination, and

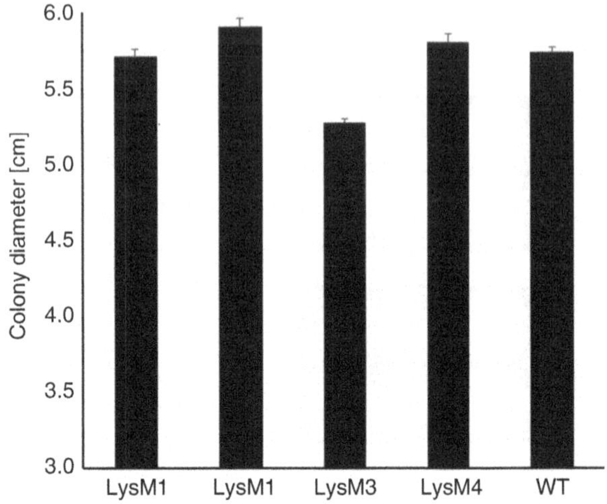

Fig. 1.3 Effect of the targeted deletion LysM genes on the growth rate. Diameter of colonies of LysM null mutants *PeLysM1, PeLysM2, PeLysM3, PeLysM4* and wild type (WT) grown on PDA medium, measured 7 days post inoculation. The results are an average of three independent measurements

effects on pathogenicity. No significant effect of the null mutations on infection incidence or the rate of lesion development was observed on 'Golden Delecius' apples. The *PeLysM3* null mutant, however, exhibited a lower rate of radial growth on PDA plates compared to the WT strain and the other mutants (Fig. 1.3). The *PeLysM3* null mutant also exhibited a significantly lower percent of spore germination, as well as shorter germ tubes on spores that did germinate. Collectively, these results suggest that *P. expansum* LysM proteins may play a potential role in growth processes, similar to TAL6 in *T. atroviride,* rather than acting as effectors in fungal-plant interactions. Notably, a yeast two-hybrid (Y2H) analysis, conducted to gain insight into the potential interaction of the secreted LysM proteins with apple proteins, revealed LysM binding to a wide range of apple proteins, including those involved in plant defense processes and iron sequestration (Levin et al. 2017). This suggests that LysM proteins in *P. expansum* may play a potential role in modifying host response to the pathogen during the infection process. Importantly, it is also possible that interruption of a single effector gene would not have a determinable effect on virulence due to complementation by other LysM family members. In this regard, among three potential LysM effectors (Mg1LysM, Mg3LysM, and MgxLysM) produced by *Mycosphaerella graminicola*, the null mutant of *Mg3LysM* affected pathogen infection, while the knockout strain of *Mg1LysM* had no effect on *M. graminicola* pathogenicity (Marshall et al. 2011). Moreover, an 80-lineage-specific LysM gene in *Verticillium dahlia* (strain VdLs17), was reported to contribute to the virulence of *V. dahliae* in tomato but not in *Arabidopsis* and tobacco (Kombrink et al. 2017). An elaborate study of the LysM protein family in *P. expansum* was recently conducted by Chen et al. (2020). The effect of 16 LysM deletion mutants on the growth and virulence of *P. expansum* on apples was analyzed. Growth rates of 3 out of 16 null mutants on a PDA medium were demonstrated to be slower, relative to the WT strain. Eleven of the 16 LysM null mutants, however, exhibited enhanced virulence on apples. Notably, only 2 of the 16 null mutants were less virulent on apples than the WT strain.

NLP Proteins in *P. expansum*

Effector proteins are generally species-specific and lack common features within and across species (Sperschneider et al. 2015). Some effector families, however, have been found to play a role in numerous plant-pathogen interactions. Nep1-like proteins (NLPs), produced by various microorganisms with different lifestyles, were shown to be involved in the virulence of several plant pathogens (Mattinen et al. 2004; Dallal Bashi et al. 2010; Santhanam et al. 2013).

Use of the effector-prediction pipeline developed by Levin et al. (2019a) identified 17 candidate genes coding for proteins most likely involved in the pathogenicity and virulence of *P. expansum*. An NLP encoding gene, *Penlp1* (PEX2_080220), was identified as one of the top effector-candidates, since *Penlp1* encoded a small (240 aa), cysteine rich (5 cysteines) protein and was highly induced in the initial stages of apple infection. Importantly, proteins possessing an NPP1-like domain were previously demonstrated to contribute to the pathogenicity of plant pathogens (Schouten et al. 2008; Dallal Bashi et al. 2010). Further analysis of the *P. expansum* genome, revealed an additional gene, *Penlp2* (PEX2_071150) possessing a NPP-1 domain (IPR008701). *Penlp1* and *Penlp2* encode a type 1 and type 3 NLP, respectively. While *Penlp1* expression was found to be strongly induced during apple infection, the highest level of *Penlp2* expression was found in ungerminated spores (Levin et al. 2019b). Although, both proteins were found to induce necrosis when transiently expressed in *Nicotiana Benthamiana* leaves, only the deletion of *Penlp1* had an effect on the pathogenicity of *P. expansum* (Levin et al. 2019b). Apples infected with the null mutant of *Penlp1*, developed smaller lesions than apples inoculated with the WT strain and the ectopic mutant (Fig. 1.4). These results are in

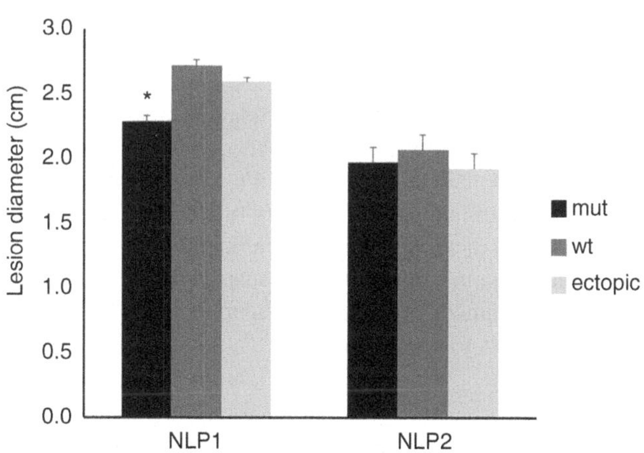

Fig. 1.4 Effect of the targeted deletion of NLP genes on *P. expansum* virulence on apple. Lesion diameter during apple infection caused by wild type (WT), ectopic mutants (E), and null mutants (mut) of *PeNLP1* and *PeNLP2* measured 6 days post inoculation. The results are an average of three independent measurements

agreement with earlier studies indicating that functional diversification exists in the superfamily of NLP1-like proteins with some members contributing to virulence while others do not (Santhanam et al. 2013).

A Y2H system was used to identify apple proteins that could potentially interact with PeNLP1. While the results obtained with the Y2H system are only preliminary, they do provide useful information for further, more comprehensive studies of the apple-*P. expansum* pathosystem. Among the 16 apple proteins identified in the Y2H system as potentially interacting with PeNLP1, three carried Bet v I and START-like domains. Bet v 1 belongs to family 10 (PR-10) plant pathogenesis-related proteins whose expression is induced by pathogen infection, wounding, or abiotic stress, and have been reported to be involved in plant defense response (Van Loon et al. 2006, Jain and Kumar 2015). Interestingly, two of the proteins (MDP0000295540, MDP0000827820) were also found to interact with the LysM motif from *P. expansum* (Levin et al. 2017). These results suggest that in the necrotroph *P. expansum,* both types of fungal effectors may be involved in modulating non-specific plant defense response. Interestingly, characterization of the transcriptomic response of apple fruit from the susceptible cultivar, 'Royal Gala', and a resistant genotype of *Malus sieversii* (PI613981) to wounding and inoculation with *P. expansum*, revealed that two major allergen Mald1 proteins (MDP0000295540 and MDP0000533638) were more highly expressed in the resistant genotype PI613981 (Ballester et al. 2017).

NLPs have been demonstrated to have an affinity for lipid membranes and to induce membrane damage (Qutob et al. 2006; Schouten et al. 2008). NLPs, however, do not act by random disruption of the phospholipid bilayers and but rather by direct injury to membrane integrity (Qutob et al. 2006). Two other apple proteins that potentially interact with PeNLP1 are MDP0000267916 and MDP0000624197, (ATPase Vo and ATPase F1 complex subunits, respectively). ATPase could potentially be a membrane target of NLP. Morikami et al. (1992) reported that mitochondrial F1-ATPase purified from several monocotyledonous plants were found to consist of five subunits, while mitochondrial F1-ATPase purified from dicotyledonous plants comprise six different subunits (Morikami et al. 1992). We speculate that the difference in ATPase subunits between monocotyledonous and dicotyledonous plants could potentially explain the fact that NLPs only induce necrosis in leaves of dicotyledonous plants. In this regard, the ability of BcNEP (an NLP from *B. cinerea*) to bind to the plasma membrane and nuclear membrane, but not chloroplast or mitochondrial membranes, in tomato and rice suspension cultures (Schouten et al. 2008) remains to be explained. Furthermore, the association of PeNLP1 with specific membranes still needs to be investigated.

PePRT Protein

Among the top 17 candidates, identified using the effector-prediction pipeline (Levin et al. 2019a), one was a proteolytic enzyme, peptidase S8 (PEX2_027670) and denoted as PePRT. Notably, PePRT had four of the requisite characteristics designated in the pipeline. It was found to have high homology to SPM1 in *M. oryzae*, a subtilisin-like serine-protease, which functions in autophagy and was demonstrated to contribute to its pathogenicity (Saitoh et al. 2009). Notably, the peptidase inhibitor I9 and peptidase S8 domains in PePRT are also present in the putative effector, AG1IA_07795 of *Rhizoctonia solani,* which induces cell death upon injection into leaves of rice, maize, and soybean. (Zheng et al. 2013).

Comparison of the null mutants of *Peprt* with the WT strain, revealed numerous differences. *Peprt* null mutants were found to have attenuated virulence on apples, a lower growth rate, less dense mycelia, and a lower number of conidiophores with shorter conidia-chains compared to the WT strain. Moreover, the process of spore germination also appeared to be affected by the deletion of *Peprt*. The hyphal structure in the *Peprt* null mutant was altered and mycelia were wider than WT mycelia (Fig. 1.5). Further, FM staining revealed that the vacuolar structure was disrupted in the mutant (Levin et al. 2019a). Consistent with these results, targeted disruption of PePRT homologs in *Aspergillus fumigatus* and in *V. dahlia* (Vd991) also resulted in

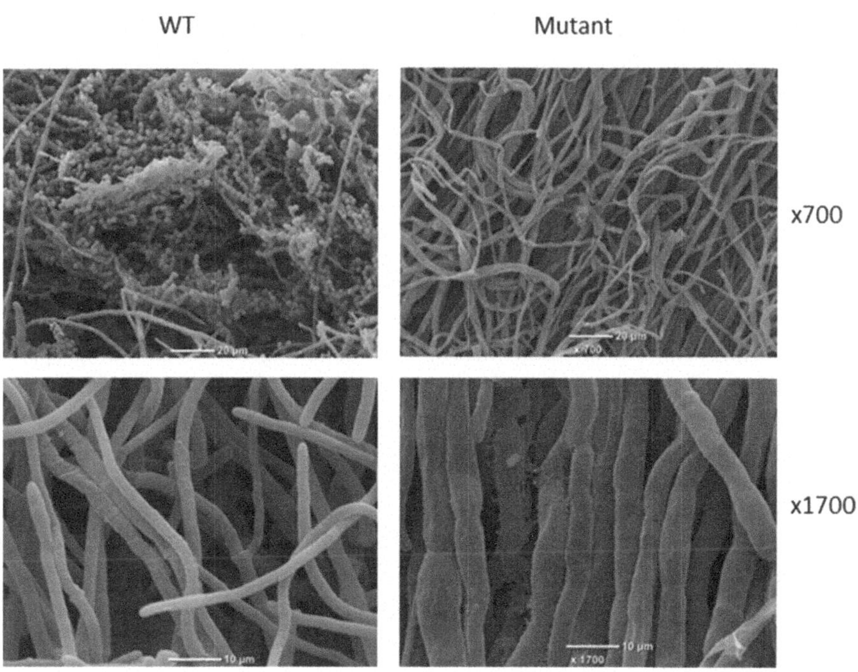

Fig. 1.5 Effect of targeted deletion of *PePRT* gene on the hyphal structure. Scanning electron microscopy image of 10 d old colony grown on PDA plates

major alterations in the sporulation process, significant reduction in growth rate, and a reduction in virulence, relative to the wild-type strains (Reichard et al. 2000; He et al. 2015). Even though PePRT appears to play a functional role in the virulence of *P. expansum* on apples, its homology to autophagy-related proteases, combined with the significant effect of the null mutation on cell functions, suggests that PePRT may potentially have a dual role in both autophagy and as a virulence factor.

Summary

P. expansum, an important postharvest pathogen on a variety of fruits, uses different pathogenicity factors to infect and develop decay in host tissue. To facilitate infection and colonization, the manipulation of host defense mechanisms at the initial stages of the infection process is crucial. In addition to the possible involvement of secondary metabolites and cell wall degrading enzymes in *P. expansum* pathogenicity on fruits, effector proteins may play an important role at the initial stages of infection.

The results of an extensive work that was undertaken to identify and characterize the role of effector proteins in *P. expansum* pathogenicity showed that the fungus is able to suppress ROS production in the stages of spore germination and colonization of fruit tissue. Down-regulation of host defense responses is attributed to the ability of *P. expansum* to produce effector proteins. The role of small cysteine-rich proteins, *Pescr1* and *Pescr2*, LysM proteins, NLPs, and one proteolytic enzyme, peptidase S8, in the virulence of *P. expansum* was also examined.

Knockout analysis of genes encoding these effector proteins revealed the possible involvement of some of them in virulence rather than pathogenicity. Yeast two-hybrid (Y2H) analysis showed that LysM binding to a wide range of apple proteins, including those involved in plant defense processes and iron sequestration. This suggests that LysM proteins in *P. expansum* may play a potential role in modifying host response to the pathogen during the infection process.

Two NLPs were identified in *P. expansum* to have a possible role in the infection on apple fruit. *Penlp1* expression was found to be strongly induced during apple infection, the highest level of *Penlp2* expression was found in ungerminated spores. Although, both proteins were found to induce necrosis when transiently expressed in *N. benthamiana* leaves, only the deletion of *Penlp1* had any effect on the virulence of *P. expansum*.

The deletion of *Peprt* affected virulence of *P. expansum* on apples and resulted in less dense mycelia, and a lower number of conidiophores with shorter conidia-chains in the culture growing *in vitro*. The process of spore germination also appeared to be affected by the deletion of *Peprt*. PePRT appears to play a role in the virulence of *P. expansum* and has autophagy-related proteases properties. This may indicate that PePRT may potentially have a dual role in both autophagy and as a virulence factor.

Collectively all the findings on the potential role of effector proteins in virulence of *P. expansum* on apples suggest that interruption of a single effector gene would not have a determinable effect on virulence due to the large repertoire of pathogenicity and virulence factors of the pathogen. Future studies may require targeting transcription factors controlling gene clusters in which several of the effector genes are located. Gene editing technologies may also hold promise in the interruption of several effector genes to ultimately determine their role in pathogenicity and virulence.

References

Abeysekara NS, Friesen TL, Keller B, Faris JD (2009) Identification and characterization of a novel host–toxin interaction in the wheat *Stagonospora nodorum* pathosystem. Theor Appl Genet 120:117–126

Alkan N, Espeso EA, Prusky D (2013) Virulence regulation of phytopathogenic fungi by pH. Antioxid Redox Sign 19:1012–1025

Ballester AR, Marcet-Houben M, Levin E, Sela N, Selma-Lázaro C, Carmona L, Wisniewski M, Droby S, González-Candelas L, Gabaldón T (2015) Genome, transcriptome, and functional analyses of *Penicillium expansum* provide new insights into secondary metabolism and pathogenicity. Mol Plant-Microbe Interact 28:232–248

Ballester AR, Norelli J, Burchard E, Abdelfattah A, Levin E, González-Candelas L, Droby S, Wisniewski M (2017) Transcriptomic response of resistant (PI613981–*Malus sieversii*) and susceptible ("Royal Gala") genotypes of apple to blue mold (*Penicillium expansum*) infection. Front Plant Sci 8:1981

Barad S, Horowitz SB, Kobiler I, Sherman A, Prusky D (2014) Accumulation of the mycotoxin patulin in the presence of gluconic acid contributes to pathogenicity of *Penicillium expansum*. Mol Plant-Microbe Interact 27:66–77

Boller T, He SY (2009) Innate immunity in plants: an arms race between pattern recognition receptors in plants and effectors in microbial pathogens. Science 324:742–744

Bolton MD, Van Esse HP, Vossen JH, De Jonge R, Stergiopoulos I, Stulemeijer IJ, Van Den Berg GC, Borrás-Hidalgo O, Dekker HL, De Koster CG, De Wit PJ (2008) The novel *Cladosporium fulvum* lysin motif effector Ecp6 is a virulence factor with orthologues in other fungal species. Mol Microbiol 69:119–136

Buron-Moles G, Torres R, Teixidó N, Usall J, Vilanova L, Viñas I (2015) Characterization of H_2O_2 production to study compatible and non-host pathogen interactions in orange and apple fruit at different maturity stages. Postharvest Biol Tec 99:27–36

Cappellini RA, Ceponis MJ, Lightner GW (1987) Disorders in apple and pear shipments to the New York market, 1972–1984. Plant Dis 71:852–856

Chen D, Li G, Liu J, Wisniewski M, Droby S, Levin E, Huang S, Liu Y (2020) Multiple transcriptomic analyses and characterization of pathogen-related core effectors and LysM family members revealed their differential roles in fungal growth and pathogenicity in *Penicillium expansum*. Mol. Genet Genomics. https://doi.org/10.1007/s00438-020-01710-9

Chisholm ST, Coaker G, Day B, Staskawicz BJ (2006) Host-microbe interactions: shaping the evolution of the plant immune response. Cell 124:803–814

Dallal Bashi ZAFER, Hegedus DD, Buchwaldt L, Rimmer SR, Borhan MH (2010) Expression and regulation of *Sclerotinia sclerotiorum* necrosis and ethylene-inducing peptides (NEPs). Mol Plant Pathol 11:43–53

De Jonge R, Van Esse HP, Kombrink A, Shinya T, Desaki Y, Bours R, Van Der Krol S, Shibuya N, Joosten MH, Thomma BP (2010) Conserved fungal LysM effector Ecp6 prevents chitin-triggered immunity in plants. Science 329:953–955

De Wit PJ, Mehrabi R, Van den Burg HA, Stergiopoulos I (2009) Fungal effector proteins: past, present and future. Mol Plant Pathol 10:735–747

Doehlemann G, Van Der Linde K, Aßmann D, Schwammbach D, Hof A, Mohanty A, Jackson D, Kahmann R (2009) Pep1, a secreted effector protein of *Ustilago maydis*, is required for successful invasion of plant cells. PLoS Pathog 5:e1000290

Droby S, Eick A, Macarisin D, Cohen L, Rafael G, Stange R, McColum G, Dudai N, Nasser A, Wisniewski M, Shapira R (2008) Role of citrus volatiles in host recognition, germination and growth of *Penicillium digitatum* and *Penicillium italicum*. Postharvest Biol Tec 49:386–396

Droby S, Wisniewski M, Macarisin D, Wilson C (2009) Twenty years of postharvest biocontrol research: is it time for a new paradigm? Postharvest Biol Tec 52:137–145

Friesen TL, Meinhardt SW, Faris JD (2007) The *Stagonospora nodorum*-wheat pathosystem involves multiple proteinaceous host-selective toxins and corresponding host sensitivity genes that interact in an inverse gene-for-gene manner. Plant J 51:681–692

Friesen TL, Faris JD, Solomon PS, Oliver RP (2008) Host-specific toxins: effectors of necrotrophic pathogenicity. Cell Microbiol 10:1421–1428

Guyon K, Balagué C, Roby D, Raffaele S (2014) Secretome analysis reveals effector candidates associated with broad host range necrotrophy in the fungal plant pathogen *Sclerotinia sclerotiorum*. BMC Genomics 15:336

He XJ, Li XL, Li YZ (2015) Disruption of Cerevisin via *Agrobacterium tumefaciens*-mediated transformation affects microsclerotia formation and virulence of *Verticillium dahliae*. Plant Pathol 64:1157–1167

Houterman PM, Cornelissen BJ, Rep M (2008) Suppression of plant resistance gene-based immunity by a fungal effector. PLoS Pathog 4:e1000061

Houterman PM, Ma L, Van Ooijen G, De Vroomen MJ, Cornelissen BJ, Takken FL, Rep M (2009) The effector protein Avr2 of the xylem-colonizing fungus *Fusarium oxysporum* activates the tomato resistance protein I-2 intracellularly. Plant J 58:970–978

Jackson AO, Taylor CB (1996) Plant-microbe interactions: life and death at the interface. Plant Cell 8:1651

Jain S, Kumar A (2015) The pathogenesis-related class 10 proteins in plant defense against biotic and abiotic stresses. Adv Plants Agric Res 3:00077

Janick J, Blase MG, Johnson DL, Jolliff GD, Myers RL (1996) Diversifying U.S. crop production. CAST, Council for Agricultural Science and Technology

Jones JD, Dangl JL (2006) The plant immune system. Nature 444:323–329

Jurick WM, Janisiewicz WJ, Saftner RA, Vico I, Gaskins VL, Park E, Forsline PL, Fazio G, Conway WS (2011) Identification of wild apple germplasm (*Malus* spp.) accessions with resistance to the postharvest decay pathogens *Penicillium expansum* and *Colletotrichum acutatum*. Plant Breed 130:481–486

Kamoun S (2006) A catalogue of the effector secretome of plant pathogenic oomycetes. Annu Rev Phytopathol 44:41–60

Kleemann J, Rincon-Rivera LJ, Takahara H, Neumann U, van Themaat EVL, van der Does HC, Hacquard S, Stüber K, Will I, Schmalenbach W, Schmelzer E (2012) Sequential delivery of host-induced virulence effectors by appressoria and intracellular hyphae of the phytopathogen *Colletotrichum higginsianum*. PLoS Pathog 8:e1002643

Kombrink A, Rovenich H, Shi-Kunne X, Rojas-Padilla E, Van den Berg GC, Domazakis E, De Jonge R, Valkenburg DJ, Sánchez-Vallet A, Seidl MF, Thomma BP (2017) *Verticillium dahliae* LysM effectors differentially contribute to virulence on plant hosts. Mol Plant Pathol 18:596–608

Lamb C, Dixon RA (1997) The oxidative burst in plant disease resistance. Annu Rev Plant Bio 48:251–275

Laugé R, Joosten MH, Van den Ackerveken GF, Van den Broek HW, De Wit PJ (1997) The in planta-produced extracellular proteins ECP1 and ECP2 of *Cladosporium fulv*um are virulence factors. Mol Plant-Microbe Interact 10:725–734

Levin E, Ballester AR, Raphael G, Feigenberg O, Liu Y, Norelli J, Gonzalez-Candelas L, Ma J, Dardick C, Wisniewski M, Droby S (2017) Identification and characterization of LysM effectors in *Penicillium expansum*. PLoS One 12:e0186023

Levin E, Kishore A, Ballester AR, Raphael G, Feigenberg O, Liu Y, Norelli J, Gonzalez-Candelas L, Wisniewski M, Droby S (2019a) Identification of pathogenicity-related genes and the role of a subtilisin-related peptidase S8 (PePRT) in authophagy and virulence of *Penicillium expansum* on apples. Postharvest Biol Tec 149:209–220

Levin E, Raphael G, Ma J, Ballester AR, Feygenberg O, Norelli J, Aly R, Gonzalez-Candelas L, Wisniewski M, Droby S (2019b) Identification and functional analysis of NLP-encoding genes from the postharvest pathogen *Penicillium expansum*. Microorganisms 7:175

Li B, Zong Y, Du Z, Chen Y, Zhang Z, Qin G, Zhao W, Tian S (2015) Genomic characterization reveals insights into patulin biosynthesis and pathogenicity in *Penicillium* species. Mol Plant-Microbe Interact 28:635–647

Liu Z, Faris JD, Oliver RP, Tan KC, Solomon PS, McDonald MC, McDonald BA, Nunez A, Lu S, Rasmussen JB, Friesen TL (2009) SnTox3 acts in effector-triggered susceptibility to induce disease on wheat carrying the *Snn3* gene. PLoS Pathog 5:e1000581

Liu Z, Zhang Z, Faris JD, Oliver RP, Syme R, McDonald MC, McDonald BA, Solomon PS, Lu S, Shelver WL, Xu S (2012) The cysteine-rich necrotrophic effector SnTox1 produced by *Stagonospora nodorum* triggers susceptibility of wheat lines harboring *Snn1*. PLoS Pathog 8:e1002467

Manning VA, Hardison LK, Ciuffetti LM (2007) Ptr ToxA interacts with a chloroplast-localized protein. Mol Plant-Microbe Interact 20:168–177

Marshall R, Kombrink A, Motteram J, Loza-Reyes E, Lucas J, Hammond-Kosack KE, Thomma BP, Rudd JJ (2011) Analysis of two in planta expressed LysM effector homologs from the fungus *Mycosphaerella graminicola* reveals novel functional properties and varying contributions to virulence on wheat. Plant Physiol 156:756–769

Mattinen L, Tshuikina M, Mäe A, Pirhonen M (2004) Identification and characterization of Nip, necrosis-inducing virulence protein of *Erwinia carotovora* subsp. *carotovora*. Mol Plant-Microbe Interact 17:1366–1375

McCallum JL, Tsao R, Zhou T (2002) Factors affecting patulin production by *Penicillium expansum*. J Food Prot 65:1937–1942

Menne KM, Hermjakob H, Apweiler R (2000) A comparison of signal sequence prediction methods using a test set of signal peptides. Bioinformatics 16:741–742

Mentlak TA, Kombrink A, Shinya T, Ryder LS, Otomo I, Saitoh H, Terauchi R, Nishizawa Y, Shibuya N, Thomma BP, Talbot NJ (2012) Effector-mediated suppression of chitin-triggered immunity by *Magnaporthe oryzae* is necessary for rice blast disease. Plant Cell 24:322–335

Morikami A, Aiso K, Asahi T, Nakamura K (1992) The delta'-subunit of higher plant six-subunit mitochondrial F1-ATPase is homologous to the delta-subunit of animal mitochondrial F1-ATPase. J Biol Chem 267:72–76

Nielsen H (2017) Predicting secretory proteins with SignalP. In: Protein function prediction (Kihara D eds). Humana Press, New York, pp 59–73

Norelli JL, Wisniewski M, Fazio G, Burchard E, Gutierrez B, Levin E, Droby S (2017) Genotyping-by-sequencing markers facilitate the identification of quantitative trait loci controlling resistance to *Penicillium expansum* in *Malus sieversii*. PLoS One 12:e0172949

Nürnberger T, Kemmerling B (2009) PAMP-triggered basal immunity in plants. Adv Bot Res 51:1–38

Osbourn AE (1996) Preformed antimicrobial compounds and plant defense against fungal attack. Plant Cell 8:1821

Pelser PDT, Eckert JW (1977) Constituents of orange juice that stimulate the germination of conidia of *Penicillium digitatum*. Phytopathology 67:747–754

Perez-Nadales E, Nogueira MFA, Baldin C, Castanheira S, El Ghalid M, Grund E, Lengeler K, Marchegiani E, Mehrotra PV, Moretti M, Naik V (2014) Fungal model systems and the elucidation of pathogenicity determinants. Fungal Genet Biol 70:42–67

Prusky D, McEvoy JL, Saftner R, Conway WS, Jones R (2004) Relationship between host acidification and virulence of *Penicillium* spp. on apple and citrus fruit. Phytopathology 94:44–51

Qin G, Tian S, Chan Z, Li B (2007) Crucial role of antioxidant proteins and hydrolytic enzymes in pathogenicity of *Penicillium expansum*: analysis based on proteomics approach. Mol Cell Proteomics 6:425–438

Qutob D, Kemmerling B, Brunner F, Küfner I, Engelhardt S, Gust AA, Luberacki B, Seitz HU, Stahl D, Rauhut T, Glawischnig E (2006) Phytotoxicity and innate immune responses induced by Nep1-like proteins. Plant Cell 18:3721–3744

Raffaele S, Win J, Cano LM, Kamoun S (2010) Analyses of genome architecture and gene expression reveal novel candidate virulence factors in the secretome of *Phytophthora infestans*. BMC Genomics 11:637

Reichard U, Cole GT, Hill TW, Rüchel R, Monod M (2000) Molecular characterization and influence on fungal development of ALP2, a novel serine proteinase from *Aspergillus fumigatus*. Int J Med Microbiol 290:549–558

Rep M (2005) Small proteins of plant-pathogenic fungi secreted during host colonization. FEMS Microbiol Lett 253:19–27

Rep M, Van Der Does HC, Meijer M, Van Wijk R, Houterman PM, Dekker HL, De Koster CG, Cornelissen BJ (2004) A small, cysteine-rich protein secreted by *Fusarium oxysporum* during colonization of xylem vessels is required for I-3-mediated resistance in tomato. Mol Microbiol 53:1373–1383

Saitoh H, Fujisawa S, Ito A, Mitsuoka C, Berberich T, Tosa Y, Asakura M, Takano Y, Terauchi R (2009) *SPM1* encoding a vacuole-localized protease is required for infection-related autophagy of the rice blast fungus *Magnaporthe oryzae*. FEMS Microbiol Lett 300:115–121

Sánchez-Torres P, González-Candelas L (2003) Isolation and characterization of genes differentially expressed during the interaction between apple fruit and *Penicillium expansum*. Mol Plant Pathol 4:447–457

Sánchez-Torres P, Vilanova L, Ballester AR, López-Pérez M, Teixidó N, Viñas I, Usall J, González-Candelas L, Torres R (2018) Unraveling the contribution of the *Penicillium expansum* PeSte12 transcription factor to virulence during apple fruit infection. Food Microbiol 69:123–135

Santhanam P, Van Esse HP, Albert I, Faino L, Nürnberger T, Thomma BP (2013) Evidence for functional diversification within a fungal NEP1-like protein family. Mol Plant-Microbe Interact 26:278–286

Sanzani SM, Reverberi M, Punelli M, Ippolito A, Fanelli C (2012) Study on the role of patulin on pathogenicity and virulence of *Penicillium expansum*. Int J Food Microbiol 153:323–331

Schneider G, Fechner U (2004) Advances in the prediction of protein targeting signals. Proteomics 4:1571–1580

Schouten A, Van Baarlen P, Van Kan JA (2008) Phytotoxic Nep1-like proteins from the necrotrophic fungus *Botrytis cinerea* associate with membranes and the nucleus of plant cells. New Phytol 177:493–505

Seidl-Seiboth V, Zach S, Frischmann A, Spadiut O, Dietzsch C, Herwig C, Ruth C, Rodler A, Jungbauer A, Kubicek CP (2013) Spore germination of *Trichoderma atroviride* is inhibited by its Lys M protein TAL 6. FEBS J 280:1226–1236

Snini SP, Tannous J, Heuillard P, Bailly S, Lippi Y, Zehraoui E, Barreau C, Oswald IP, Puel O (2016) Patulin is a cultivar-dependent aggressiveness factor favouring the colonization of apples by *Penicillium expansum*. Mol Plant Pathol 17:920–930

Sperschneider J, Gardiner DM, Taylor JM, Hane JK, Singh KB, Manners JM (2013) A comparative hidden Markov model analysis pipeline identifies proteins characteristic of cereal-infecting fungi. BMC Genomics 14:807

Sperschneider J, Ying H, Dodds PN, Gardiner DM, Upadhyaya NM, Singh KB, Manners JM, Taylor JM (2014) Diversifying selection in the wheat stem rust fungus acts predominantly on pathogen-associated gene families and reveals candidate effectors. Front Plant Sci 5:372

Sperschneider J, Dodds PN, Gardiner DM, Manners JM, Singh KB, Taylor JM (2015) Advances and challenges in computational prediction of effectors from plant pathogenic fungi. PLoS Pathog 11:e1004806

Syme RA, Hane JK, Friesen TL, Oliver RP (2013) Resequencing and comparative genomics of *Stagonospora nodorum*: sectional gene absence and effector discovery. G3 (Bethesda) 3:959–969

Tahir I, Nybom H, Ahmadi-Afzadi M, Røen K, Sehic J, Røen D (2015) Susceptibility to blue mold caused by *Penicillium expansum* in apple cultivars adapted to a cool climate. Eur J Hortic Sci 80:117–127

Takahara H, Hacquard S, Kombrink A, Hughes HB, Halder V, Robin GP, Hiruma K, Neumann U, Shinya T, Kombrink E, Shibuya N (2016) *Colletotrichum higginsianum* extracellular LysM proteins play dual roles in appressorial function and suppression of chitin-triggered plant immunity. New Phytol 211:1323–1337

Tan KC, Oliver RP, Solomon PS, Moffat CS (2010) Proteinaceous necrotrophic effectors in fungal virulence. Funct Plant Biol 37:907–912

Tian S, Qin G, Li B (2013) Reactive oxygen species involved in regulating fruit senescence and fungal pathogenicity. Plant Mol Biol 82:593–602

Torres MA (2010) ROS in biotic interactions. Physiol Plant 138:414–429

Torres MA, Jones JD, Dangl JL (2006) Reactive oxygen species signaling in response to pathogens. Plant Physiol 141:373–378

Torto TA, Li S, Styer A, Huitema E, Testa A, Gow NA, Van West P, Kamoun S (2003) EST mining and functional expression assays identify extracellular effector proteins from the plant pathogen *Phytophthora*. Genome Res 13:1675–1685

Van Kan JA (2006) Licensed to kill: the lifestyle of a necrotrophic plant pathogen. Trends Plant Sci 11:247–253

Van Loon LC, Rep M, Peiterse CM (2006) Significance of inducible defense-related proteins in infected plants. Annu Rev Phytopathol 44:135–162

Vilanova L, Viñas I, Torres R, Usall J, Buron-Moles G, Teixidó N (2014) Acidification of apple and orange hosts by *Penicillium digitatum* and *Penicillium expansum*. Int J Food Microbiol 178:39–49

Visagie CM, Houbraken J, Frisvad JC, Hong SB, Klaassen CHW, Perrone G, Seifert KA, Varga J, Yaguchi T, Samson RA (2014) Identification and nomenclature of the genus *Penicillium*. Stud Mycol 78:343–371

Wang X, Jiang N, Liu J, Liu W, Wang GL (2014) The role of effectors and host immunity in plant–necrotrophic fungal interactions. Virulence 5:722–732

Weiberg A, Wang M, Lin FM, Zhao H, Zhang Z, Kaloshian I, Huang HD, Jin H (2013) Fungal small RNAs suppress plant immunity by hijacking host RNA interference pathways. Science 342:118–123

Wouters MFA, Speijers GJA (1996) Patulin. Food Add Ser 35:337–402

Zheng A, Lin R, Zhang D, Qin P, Xu L, Ai P, Ding L, Wang Y, Chen Y, Liu Y, Sun Z (2013) The evolution and pathogenic mechanisms of the rice sheath blight pathogen. Nat Commun 4:1–10

Chapter 2
Insights into Fruit Defense Mechanisms Against the Main Postharvest Pathogens of Apples and Oranges

Rosario Torres, Laura Vilanova, Josep Usall, and Neus Teixidó

Abstract *Penicillium digitatum* and *P. expansum* are the most devastating pathogens of citrus and pome fruits respectively, causing significant economic losses during postharvest handling worldwide. To obtain new rational and environmentally friendly control alternatives, a better understanding of the fruit-pathogen interaction may be considered as a novel perspective for the control of postharvest diseases. The main objective of our studies was to gain insights into the fruit-pathogen interactions, specifically in oranges and apples defence responses against compatible and non-host pathogens. For such purposes, firstly a deep study of infection capacities of both pathogens in oranges and apples at different conditions was performed. Later, we characterized the effect of wound response in oranges and apples harvested at different maturity stages and temperatures, as well as we identified the possible compounds involved in the wound healing process. In addition, a visualization of lignin, suberin and callose in apple and oranges tissue was conducted using histochemical tests. Finally, in apples, a transcriptomic study in response to compatible and non-host pathogen was conducted; and the expression of several genes involved in the phenylpropanoid pathway were quantified in citrus fruit.

Keywords *P. expansum* · *P. digitatum* · Host-pathogen interaction · Lignification · Phenylpropanoid pathway · Transcriptomics analysis

---

R. Torres (✉) · L. Vilanova · J. Usall · N. Teixidó
IRTA, Postharvest Program, Edifici Fruitcentre, Parc Científic i Tecnològic Agroalimentari de Lleida, Lleida, Catalonia, Spain
e-mail: rosario.torres@irta.cat; laura.vilanova@irta.cat; josep.usall@irta.cat; neus.teixido@irta.cat

D. Spadaro et al. (eds.), *Postharvest Pathology*, Plant Pathology in the 21st Century 11, https://doi.org/10.1007/978-3-030-56530-5_2

Introduction

Oranges and apples are some of most produced fruits worldwide. The production of citrus in Mediterranean regions is 25,216 thousand tons in 2016, and Spain is the first producer with 6,882 thousand tons. Specifically, Spain contributed with 3,641 thousand tons of oranges and, therefore, remains the major contribution in Mediterranean Region with around 25% of total production (FAO 2017). In turn, as important apple producer, Spain contributed to the apple worldwide production with 473 thousand tons in 2018 (Prognosfruit 2018).

Quality deterioration of fresh fruits due to fungal pathogens is one of the main economic losses produced in postharvest handling worldwide. The most common cause of citrus and pome fruit decay is the development of rots caused by *Penicillium* spp. which are one of the most common and destructive genera in postharvest diseases, affecting a wide variety of fruit. In particular, *P. digitatum* and *P. expansum* are the most devastating postharvest pathogens of citrus and pome fruits, respectively. Although both are close related species, they show a different range of hosts. Whereas *P. digitatum* causes only green mould in citrus and could be considered as specialist species, *P. expansum* has the ability to infect a broader range of hosts (Fig. 2.1). These necrotrophic pathogens need wounds to enter tissues and start the infection process.

The traditional way to control these diseases is the usage of synthetic fungicides. However, there are important problems associated by their use, hence, actual trends are directed to find new rational and environmental friendly control alternatives focused on physical, chemical and biological strategies. These facts justify the need and the interest for more detailed studies of host-pathogen interaction to increase the knowledge on both virulence factors of pathogens and fruit defence mechanisms with the final aim to improve control strategies (Tian et al. 2016).

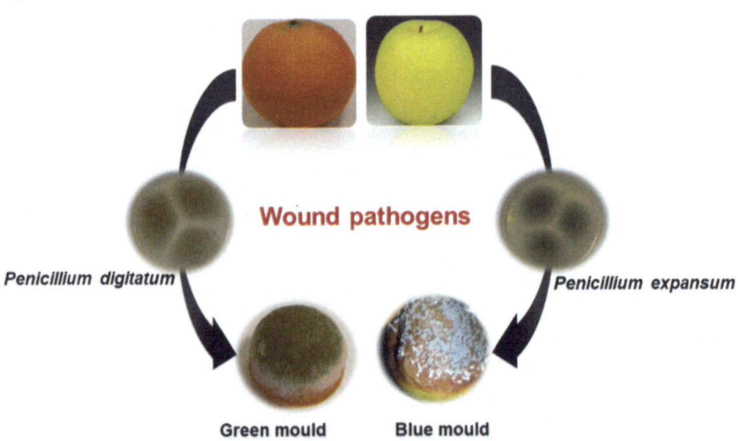

Fig. 2.1 The main postharvest pathogens of oranges and apples, *Penicillium digitatum* and *P. expansum*, respectively

At this regard, new perspectives to control postharvest diseases are needed and some of them are based on the knowledge of fruit-pathogen interaction. On one hand, the capacity of fungi to adapt to new conditions, overcoming plant defence mechanisms and becoming resistant to fungicides is huge. On the other hand, the defence machinery of the fruit is also complex and unknown, and evolved a multi-level series of structural and biochemical barriers that are both constitutive or pre-formed and inducible. Both components of host-pathogen interactions should be addressed from different perspectives, pathological, biochemical and molecular studies, to elucidate the factors that can modulate both compatible and non-host interactions.

The main proposal of this chapter is to approach the research line related to control postharvest diseases on pome and citrus fruit from new perspective, mainly focused on fruit-pathogen interactions in apple/oranges- *Penicillium* spp.

Fruit–*Penicillium* spp. Interaction

In many instances, the differences in the outcome of a host-pathogen interaction relies on a rapid and efficient deployment of defense responses, which will be modulated depending on whether the host-pathogen interaction involves a compatible or non-host interaction. For that purpose, *P. digitatum* and *P. expansum* were defined as compatible pathogens of oranges and apples, respectively, meanwhile *P. expansum* and *P. digitatum* were described as non-host pathogens in oranges and apples, respectively (Fig. 2.2).

Fruit defenses are complex and constitute a multilevel series of structural and biochemical barriers that are both constitutive and inducible. Therefore, a multidisciplinary approach including pathological, biochemical and molecular studies were conducted to unravel the underlying mechanisms related to fruit response against both postharvest pathogens. In particular, in this chapter we have focused our research in the following aspects: (1) infection capacity of each pathogen in oranges and apples; (2) characterization of fruit healing process; (3) identification and expression analysis of genes involved in fruit defence mechanisms.

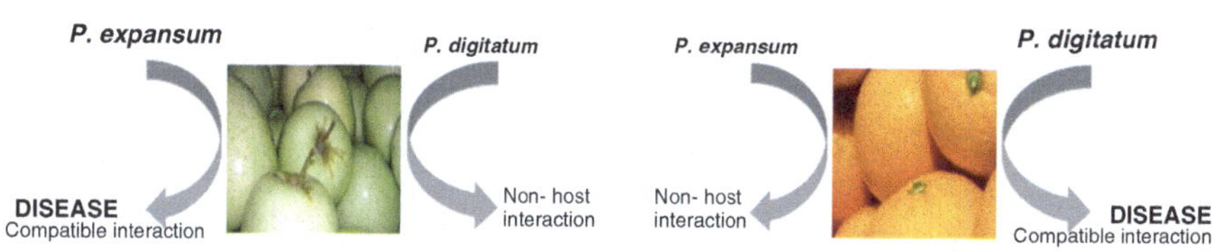

Fig. 2.2 Compatible and non-host interaction in apple and oranges

Infection Capacity of Both Pathogens on Apples and Oranges

The infection capacities of *P. digitatum* on oranges and of *P. expansum* on apples take into account several key factors such as i) maturity of fruit, ii) pathogen inoculum concentrations and iii) storage temperatures (Fig. 2.3).

Vilanova et al. (2012a, b) clearly defined the compatible and non-host interactions at both studied pathosystems. The main results showed that in compatible interactions the lowest growth rate of both pathogens was observed in the immature harvest fruit stored at both 20 °C and cold temperatures (4 °C and 0 °C for oranges and apples, respectively). In addition to that, the lower growth rate was accentuated at lower concentration of inoculum of pathogens as 10^4 conidia mL^{-1}. Several authors have observed higher susceptibility of overripe apples compared to immature apples against *P. expansum* and *Botrytis cinerea*, which highlight the role of maturity in the susceptibility of fruit to fungal diseases (Torres et al. 2003; Neri et al. 2010; Su et al. 2011). Furthermore, in oranges no differences were observed between the growth rates of *P. digitatum* in oranges stored at both 20 °C and 4 °C, at the concentration of 10^7 conidia mL^{-1} at any maturity stage. This fact could be explained because the inoculum concentration is too high that the maturity states of fruit do not affect the growth rate of this pathogen.

In apples and oranges, the storage temperature influenced the growth rate and visual symptoms of the decay of these pathogens and need shorter incubation times at 20 °C than at cold temperature (Vilanova et al. 2012a, b). These results coincide with those obtained by Baert et al. (2007a, b) in apples inoculated with different strains of *P. expansum*. Moreover, in *in vitro* studies in PDA medium, Buron-Moles et al. (2012) showed a decrease in the growth rate of *P. digitatum* and *P. expansum* at cold temperature (4 °C and 0 °C, respectively) compared to 25 °C.

Regarding inoculum concentration, in general no significant differences were observed at any concentration of the pathogen tested (from 10^4 to 10^7 conidia mL^{-1}). Immature apples inoculated with *P. expansum* and stored at 20 °C, however, showed

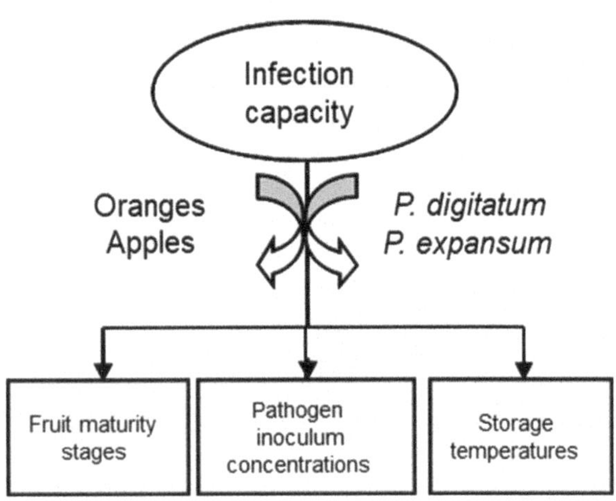

Fig. 2.3 Schematic diagram of studies on infection capacity of *P. digitatum* and *P. expansum* on oranges and apples

a different growth rate dependent on inoculum concentration, in contrast, these differences were not observed in oranges (Vilanova et al. 2012a, b). In a similar work, Morales et al. (2008) observed differences in the growth rate of *P. expansum* at different inoculum concentrations in apples stored at 20 °C.

On the other hand, considering the non-host interaction, e.g. *P. expansum*-oranges and *P. digitatum*-apples, these studies demonstrated that they became compatible interaction depending on the conditions and, therefore, they cannot strictly considered non-host pathogens. Previous studies conducted by Macarisin et al. (2007) have already shown that *P. expansum* could germinate and develop decay in citrus fruit but only at the point of inoculation. These authors only observed green mould in apples after treatment of wounds with different organic acids (citrus, ascorbic and oxalic) or with the enzyme catalase (CAT) before inoculation, pointing out the role of these substances in the suppression of the production of H_2O_2. Vilanova et al. (2012b) showed the ability of *P. expansum* to infect oranges by itself, without any pre-treatment, at 20 °C and 4 °C. However, incidence and severity of these infected oranges were higher at 4 °C than at 20 °C (Fig. 2.4a). This behaviour can be explained because wound healing, as fruit defence mechanisms, is slower at 4 °C than at 20 °C, together with the fact that *P. expansum* is a pathogen well-adapted to low temperatures. Studies carried out by Ismail and Brown (1975) already pointed out that healing of "Valencia" oranges at 5 °C was slower than at 30 °C. Later, Brown and Barmore (1983) showed that both phenolic and lignification substances could be the responsible of resistance of curing oranges against *P. digitatum* infection. In turn, the formation of lignin increases at 20 °C more than at 2.5 °C (Mulas et al. 1996).

In contrast, Vilanova et al. (2012a) demonstrated that *P. digitatum* showed some visual symptoms of decay in apples but only restricted at the site of infection (Fig. 2.4b). Only overmatured apples were decayed by the highest concentration of *P. digitatum* (10^7 conidia mL^{-1}), but exceptionally (Fig. 2.5). In addition, Buron-Moles et al. (2012) corroborated these results visualizing the infection process in apples using *P. digitatum* tagged with a green fluorescent protein (GFP). Likewise,

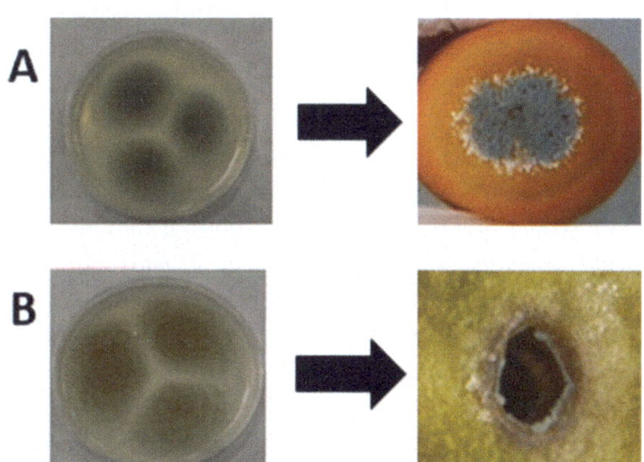

Fig. 2.4 (**a**) *P. expansum* development on oranges at 4 °C; (**b**) *P. digitatum* development on apples at 4 °C was restricted to the initial inoculation site

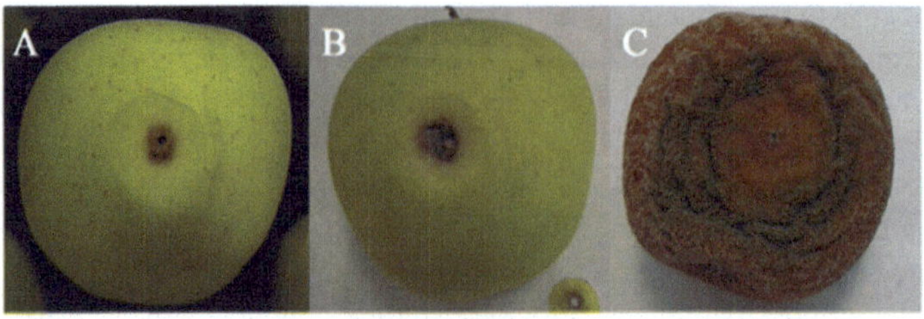

Fig. 2.5 Green mould caused by *P. digitatum* on (**a**) immature, (**b**) matures, and (**c**) overmatured apples at inoculum concentration of 10^7 conidia mL^{-1} and storage temperature of 20 °C

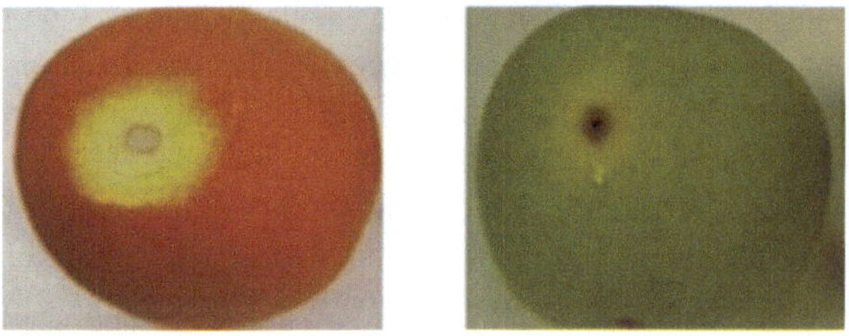

Fig. 2.6 Orange and apple inoculated with *P. expansum* and *P. digitatum*, respectively, with only visible reaction around the inoculation site

Louw and Korsten (2014) have also revealed that *P. digitatum* can infect different varieties of apples and pears. In contrast, at 0 °C, no signs of infection by *P. digitatum* were observed at any studied conditions. The overall results may explain by the ecophysiological difference between *P. digitatum* and *P. expansum* (Plaza et al. 2003; Buron-Moles et al. 2012).

A relevant fact in these non-host interactions were detected: when *P. expansum* and *P. digitatum* were not able to infect oranges and apples, an important reaction in the skin and pulp was observed (Fig. 2.6). In addition, this reaction increased proportionally to pathogen concentration and decreased as maturity advanced. This reaction could be associated to a rapid localized necrosis of non-host resistance (Mysore and Ryu 2004). Macarisin et al. (2007) in lemons also observed that approximately 4–5 days after inoculation with the non-host pathogen *P. expansum* appeared the hypersensitive response (HR), with dead cells and lignified tissue surrounding the wounds. HR usually appears in incompatible interactions (Mysore and Ryu 2004) and is recognized by brown necrotric areas resulting from a localized collapse of tissues as well as dead cells at the point of inoculation. In studies in coffee leaves (Silva et al. 2002) and melon (Romero et al. 2008), it was observed rapid HR in incompatible interactions. Whilst this response has been described to prevent the development of biotrophic fungi (Lamb and Dixon 1997; Lu and Higgins 1999), it may not prevent the consequent development of necrotrophic fungi (Mayer et al. 2001). In turn, some authors suggest that it may even stimulate the development of necrotrophic fungi (Govrin and Levine 2000).

In summary, our studies demonstrated that the inoculation of a compatible pathogen (*P. digitatum*-oranges and *P. expansum*-apples) at different inoculum concentrations always showed decay development. However, in oranges and apples, the most important differences in rot dynamics among harvests were found at the lowest inoculum concentration analysed (10^4 conidia mL^{-1}). Surprisingly, depending on the combination of factors (maturity stage, inoculum concentration and storage temperature), the non-host interaction (*P. expansum*-oranges and *P. digitatum*-apples) became compatible.

Orange and Apple Defence Mechanisms

Lack of penetration by fungi into non-host tissue can be associated with the deposition of substances or compounds in the host epidermal cell walls. Based on above results, a study of the host defence responses to abiotic (wound) and biotic (pathogen) stresses was conducted following a pathological, biochemical and molecular approach to elucidate the defence reactions underlying in these interaction (Fig. 2.7).

Wound Response Process

P. digitatum and *P. expansum* are considered as wound fungi, i.e., they need a wound on the fruit epidermis in order to start the infection (Kavanagh and Wood 1967; Spotts et al. 1998). Hence, good handling practices during harvest to avoid wounds in fruit are important to reduce postharvest losses. Beside that, a better knowledge on healing process of apples and oranges is needed.

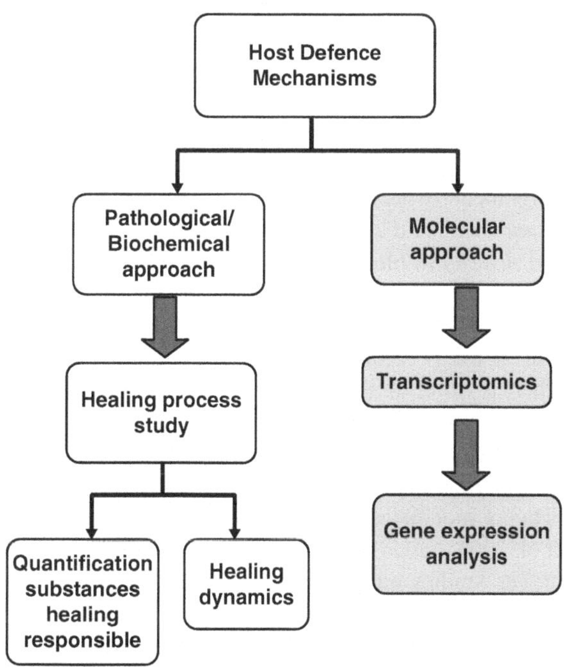

Fig. 2.7 Schematic diagram of studies on host defence mechanisms in oranges and apples

Fig. 2.8 Healing process in (**a**) oranges wounded and inoculated at different times with *P. digitatum* and (**b**) apples wounded and inoculated at different times with *P. expansum* and incubated at 20 °C and 85% RH

It is known that different treatments (curing, light UV) or inducing substances increased the resistance of fruit against post-harvest pathogens (Droby et al. 1993; Droby et al. 1999; Ballester et al. 2010, 2011; Shao et al. 2010). Vilanova et al. (2013, 2014a) studied the healing process *per se* of apples and oranges at different maturity stages when were wounded and infected after different times by *P. expansum* and *P. digitatum*. The healing process in both oranges and apples has an important effect limiting infection ability of both host pathogens when the fruit were stored at 20 °C (Fig. 2.8).

In addition, this effect was most outstanding in the immature and mature fruit than in overmatured. Regarding immature and commercial oranges, it was observed a significantly decrease in incidence and severity of wounds inoculated with *P. digitatum* 7 and 10 d after wounding, while on apples, a similar reductions in incidence and severity of blue mould it was achieved only 3 d after wounding (Vilanova et al. 2013, 2014a). Somehow, the skin of the oranges may be the responsible of this higher susceptibility to rot compared to apples. Brown et al. (1978) showed a reduction in the percentage of infected oranges by *P. digitatum* 3 d after wounding, whereas Su et al. (2011) and Shao et al. (2010) demonstrated a decrease in incidence of decayed apples by *B. cinerea* and *P. expansum*, respectively 96 h after wounding. In pears, Spotts et al. (1998) found greater resistance of fruit to *P. expansum* infection 2 d after wounding. Likewise, the greatest resistance was in green peppers inoculated with *Colletotrichum acutatum* only 1 h after wounding (Kim et al. 2008).

The studies on wounding process on oranges and apples demonstrated that it depends on temperature (Vilanova et al. 2013, 2014a). Both temperature and relative humidity are two of the most important factors affecting the process of healing

(Brown 1989). The temperature must be high enough to accelerate the development of metabolic reactions involved in the process of healing and, in turn, relative humidity should be sufficient to avoid drying or killing of tissues near damaged cells. Generally, temperatures above 10 °C and relative humidity higher than 85% are necessary to start the healing process in citrus fruit (Brown 1989). Moreover, the incidence and severity of wounded oranges inoculated with *P. digitatum* and stored at 4 °C for 30 d were lower when healing time increased. However, 45 d after inoculation, in commercial and overmatured oranges, there were no differences in the incidence between the different days of healing (Vilanova et al. 2013). This indicates that low temperatures make the healing process not effective enough to prevent infection with *P. digitatum*. In apples stored at 0 °C, the healing process was even less effective preventing infections of *P. expansum* than that observed in oranges. Lakshiminarayana et al. (1987) observed that apples inoculated with *P. expansum* or *B. cinerea* at 4 d after wounding and stored at 5 °C presented a substantial resistance to infection, although these authors considered that this healing time was scarce time to modify the cell wall to stop infection.

In contrast to these results, overmatured fruit showed few differences at different healing times, hence, once again it was clearly demonstrated the key role of maturity in susceptibility of oranges and apples to postharvest pathogen infection. Similar results were observed by Su et al. (2011) in overmatured apples inoculated with *B. cinerea* 96 h after wounding wounds: apples showed a higher incidence and severity of grey mould that immature or commercial apples. However, they did not observe differences in incidence of decay among harvest time when fruits were inoculated just after wounding.

Fruit maturity is a process that involves significant changes in the biochemical level such as alteration of the cell membrane (Cantu et al. 2008a). Consequently, maturity can modify the susceptibility of fruit by decreasing their defence responses and making them more susceptible to pathogen attack. As described above, overripe can trigger that the non-host *P. expansum* may even infect and develop decay in oranges. Some authors (Stange et al. 2002) defined that citrus skin is an inappropriate and even toxic environment for germination and growth of non-host moulds. In our study related to wound response, however, observed that *P. expansum* was able to infect and even develop rot at different times of healing in commercial and overmatured oranges at 20 °C, and at any maturity stage at 4 °C (Vilanova et al. 2013). Similar results were also observed in apples at 20 °C; *P. digitatum* was able to infect overmatured apples at 20 °C, even those inoculated 10 d after wounding, but not at 0 °C.

Cell Wall Reinforcement

Wound healing process could be associated with a limited accumulation of phenolics and lignin-like compounds in cell wall thickening around wounds and, hence, be an effective protection against pathogen infections. In our studies, fruits inoculated with the non-host pathogen (oranges inoculated with *P. expansum* and apples inoculated with *P. digitatum*) showed a visible reaction around wounds when pathogen was not able to infect them (Fig. 2.6). To identify the substances involved in this reaction, a more in–depth study was conducted. Previous authors showed a direct relation of deposition of some substances and orange resistance against fungal infections (Brown and Barmore 1983; Baudoin and Eckert 1985). Therefore, different histochemical tests to detect lignin, suberin and callose and to characterize the accumulation of substances at the site of inoculation were used. Lignin was detected using Maüle and toluidine blue O reactions, suberin by Sudan IV test, and callose by Aniline blue test (Fig. 2.9). Both tests used to detect lignin showed a positive reaction at short response times (24 h and 48 h for apples and oranges, respectively) in both compatible and non-host interactions. These results suggest that lignin production is not exclusive for compatible pathogens. Remarkably, reaction of control fruits (fruits wounded) was not equal on oranges and on apples: wounded apples showed a positive reaction in lignin test at short time, while oranges always showed a negative reaction. Therefore, lignin production seems to be related with abiotic (wounds) and biotic (pathogen) stresses, and appears more important in apples than in oranges.

Moreover, lignification observed in both oranges and apples inoculated with the non-host pathogen at long-term of response (7 d) was stronger than at 24 h or 48 h. In addition, lignification was apparently more significant in immature fruit than in commercial fruit since non-positive reaction was observed in overmatured fruit.

Neither of the tests used in Vilanova et al. (2012a, b) to detect suberin and callose showed a positive reaction either in oranges or in apples. According to Lai et al. (2003) using nuclear magnetic resonance spectroscopy (NMR spectroscopy) the substances present in curing grapefruit were suberin and not lignin. Regarding the quantification of lignin, Vilanova et al. (2013, 2014a) showed that immature control (mock inoculation at different times after wounding) in oranges and apples increased lignin content along storage period. Su et al. (2011) also observed an increase in lignin content in immature "Gala" apples compared to overmatured ones. Valentines

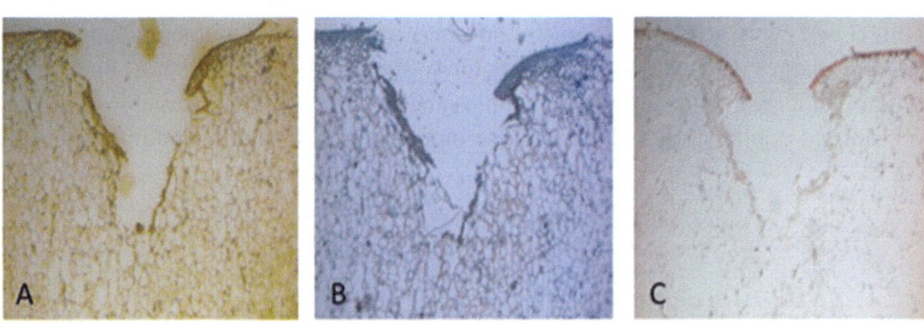

Fig. 2.9 Images of histochemical tests: (**a**) Maüle reaction, (**b**) Blue Toluidine and (**c**) Sudan IV, in immature apples inoculates inoculated with *P. digitatum* at the concentration of 10^7 conidia mL^{-1} after 7 d at 20 °C and 85% RH. Pictures (**a**) and (**b**) show a positive reaction to lignin, and (**c**) test Sudan IV shows a negative reaction to suberin

et al. (2005) also found higher lignin content in apples which denoted a major resistant to *P. expansum* infection. Overall results suggest that lignin has a role in apple defence against *P. expansum*.

Immature oranges and apples inoculated with the non-host pathogen showed an important increase in lignin content 7 d and 72 h after inoculation, respectively. These results are in agreement with histochemical results obtained by Vilanova et al. (2012a, b), in which the most intense reaction was observed in immature fruits 7 d after inoculation. In the non-host interactions, lignification could begin few days after infection; however, its accumulation was more remarkable at long-term response. In addition, the highest accumulation of lignin was observed in immature fruit that usually are not infected.

Regarding the lignin content in compatible interaction, our studies showed unusually high absorbance values. These values obtained with the acetyl bromine method may be attributed to an increase in polysaccharides degradation and suggests that other factors than solubilisation of lignin were responsible for the increase in absorbance (Hatfield et al. 1999). When necrotrophic pathogens infect fruits, they secrete various cell wall degrading enzymes that increase the polysaccharides (Cantu et al. 2008b). Therefore, the acetyl bromide method applied to macerated tissues could produce erroneous absorbance values.

Molecular Studies in Fruit-*Penicillium* Interactions

The availability of the complete genome sequence of fruit has greatly aided the development of "omic" approaches such as genomics, transcriptomics and proteomics. Specifically, in *Malus* x *domestica*, the information gained from those studies could be particularly useful in the field of postharvest pathology. In our studies, a transcriptomic analysis of apple to understand fruit disease resistance in response to compatible (*P. expansum*) and non-host (*P. digitatum*) pathogens was conducted (Vilanova et al. 2014b). Moreover, the expression of several genes involved in the phenylpropanoid pathway were quantified to define their role in citrus fruit response against pathogen and non-host pathogen (Vilanova et al. 2013).

Transcriptomic Approach in Apples

The first transcriptomics study to elucidate the genes involved in the fruit development of "Fuji" apples was conducted by Lee et al. (2007) using a microarray containing 6,253 cDNAs. Few years later, Soria-Guerra et al. (2011) developed a microarray containing approximately 40,000 probes, including positive and negative controls, from 34 libraries of cDNA constructed from vegetative and reproductive apple tissues at different developmental stages, different genotypes and different types of biotic and abiotic stresses. This microarray was used and validated in different studies: in one of them the aim was to characterize the gene expression along fruit development of different varieties of apple (Soria-Guerra et al. 2011); in another one the goal was to study the response of apples after the application of an abscision-promoting treatment (Zhu et al. 2011); in the last ones the objectives were to investigate the response of apples to *Erwinia amylovora* (Bocsanczy et al. 2009; Sarowar et al. 2011). The microarray was also used in our study (Vilanova et al. 2014b) to perform a global analysis of genes involved in response of immature apples to a compatible pathogen (*P. expansum*) and a non-host pathogen (*P. digitatum*) at 20 °C (Fig. 2.10). The results of this study indicated a different pattern of apples in response to *P. expansum* and *P. digitatum* infection. Functional annotation based on gene ontology terms (GO) showed that biological processes of apples significantly modified in response to *P. expansum* were related to defence responses and oxidative stress, meanwhile to *P. digitatum* were related to phenylpropanoid pathway. Our microarray study identified 3,770 differentially expressed probes that could be used to further study of apple-*Penicillium* spp.

One of the first defence responses detected in a host-pathogen interaction is the oxidative burst that is characterized by a significant increase in ROS levels (Levine et al. 1994; Low and Merida 1996). Plant nicotinamide adenine dinucleotide phosphate oxidases, also called "Respiratory burst oxidases homologues (Rbo)" was described as an important source of ROS species in majority of plant–pathogen interactions (Torres et al. 2006). ROS metabolism is controlled by Rbos, and other enzymes including superoxide dismutase (SOD), catalase (CAT), and ascorbate peroxidase (APX) (De Gara et al. 2003). SOD activity plays a role in the dismutation of superoxide radicals, while CAT and APX activity contribute to the elimination of hydrogen peroxide (H_2O_2) (Levine et al. 1994). Our microarray data showed

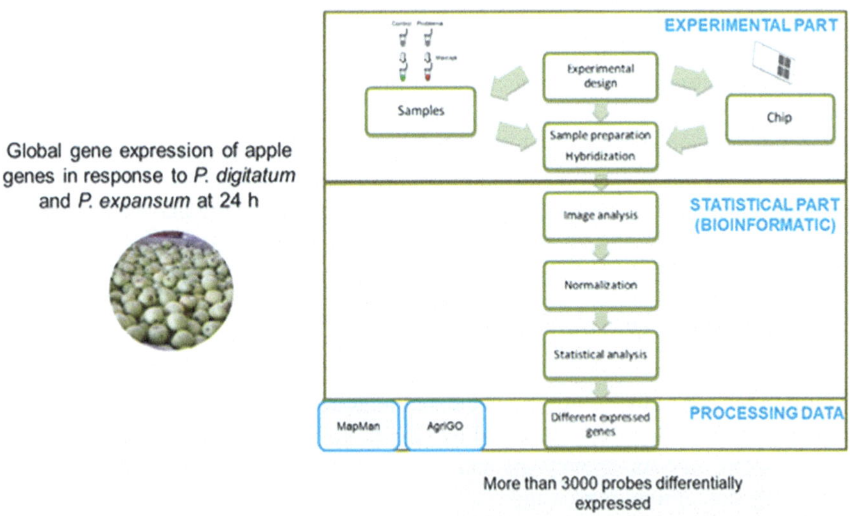

Fig. 2.10 General scheme to conduct the global study in apple in response to *P. expansum* (pathogen) and *P. digitatum* (non-host pathogen)

an induction of the expression of the *Rbo* gene in apples inoculated with *P. digitatum* compared to apples inoculated with *P. expansum*. First, this fact pointed out that an increase in ROS content could be related to defence response of apples against *P. digitatum*. On the other hand, apples inoculated with *P. expansum* showed a greater increase in the expression of genes encoding ROS-detoxifying enzymes such as SOD, APX and POX (peroxidase) compared to apples inoculated with *P. digitatum*. Nevertheless, there was no increase in the expression of the *Rbo* gene when the apples were inoculated by *P. expansum*, indicating that the highest level of *Rbo* expression may be earlier than 24 h. Torres et al. (2003) showed that the largest production of H_2O_2 in apples inoculated with *P. expansum* awas detected approximately 6 h after inoculation. This ability of *P. expansum* to prevent the oxidative burst, suppressing the production of H_2O_2 in apples 24 h after inoculation, may be closely related to its pathogenicity. These results are in agreement to those obtained by Macarisin et al. (2007) in oranges where they observed that *P. digitatum* (compatible pathogen) suppressed the oxidative burst on host cell, meanwhile *P. expansum* (non-host pathogen) triggered an accumulation of H_2O_2. The relationship between ROS levels and detoxifying enzyme levels is considered to be a key factor in determining the H_2O_2 and O_2^- tissue-levels (Mittler et al. 1999). Ballester et al. (2006) suggested an imbalance among the activities of enzymes that generate and degrade H_2O_2 to promote the infection process, as has been reporter for the necrotrophic fungi *B. cinerea* and *Sclerotinia sclerotiorum* (Govrin and Levine 2000). Pathogenicity of both moulds seems to be dependent on ROS levels generated during the infection because they may benefit from cell necrosis caused by a hypersensitive response (HR) (Govrin and Levine 2000). Additionally, the *bcsod1* gene encoding for a Cu-Zn SOD has been described in *B. cinerea* as a virulence factor (Rolke et al. 2004). Other study also shows that the virulence factor of *Monilinia fructicola* MfCUT1 is induced in the presence of ROS generated during infection process (Chiu et al. 2013).

The microarray data also indicated a change in the gene expression of heat shock proteins (HSPs). HSPs are one class of chaperones produced in response not only to high temperatures if not to stress, to exposure to heavy metals, pathogens attack and disturbance in intracellular calcium levels (Vierling 1991). In our study, we observed an induction in the *HSP101* gene expression in apples inoculated with *P. expansum* and in the *HSP70* gene expression in apples inoculated with *P. expansum* and *P. digitatum* (Vilanova et al. 2014b). These results indicated that *HSP101* might be more specific for pathogen compatible although *HSP70* may be related to the pathogen attack. Little is known about the role of HSPs in response to pathogens because most of the HSP studies have been focused on plant abiotic stresses. In fruits, for example, the expression HSP has been characterized in response to heat treatments used for control post-harvest diseases (Lauxmann et al. 2012; Pavez et al. 2013). Specifically, Yun et al. (2013) and Pavoncello et al. (2001) observed an induction of genes encoding HSP in heat-treated citrus fruit related with an increase of resistance to *P. italicum* and *P. digitatum*.

A differential expression in other genes related to defence-related proteins was observed, as genes encode for pathogenesis-related proteins (PR-proteins), and other related to synthesis of callose (Vilanova et al. 2014b). The PR proteins are divided into 17 families, including glucanases (GLU), chitinases (CHI), taumatins (TAU), peroxidase (POX) and defensins (DEFL) (van Loon et al. 2006; van Loon and van Strien 1999). The results showed that the inoculation of apples with *P. expansum* induced more PR proteins than *P. digitatum*, including two CHI, one endoglucanase (EGL), a TAU and a DEFL proteins. The outcomes indicated that induction of genes encoding PR proteins are related to fruit defence against pathogenic fungi. Concerning fruit, there is scarce information on induction of PR genes in response to pathogens. Most of the studies are based on the induction of PR proteins in response to PR proteins biological control and abiotic stress conditions. Ballester et al. (2010) correlated an increase in CHI and GLU enzymes activity with upregulation of genes encoding these enzymes when oranges are firstly cured and after that inoculated with *P. digitatum*. On the other hand, in oranges, Hershkovitz et al. (2012) detected an induction of the CHI and in apples, Quaglia et al. (2011) detected an induction of the CHI, GLU and TAU when the fruits were treated with biocontrol agents. Buron-Moles et al. (2014) in proteomic studies, revealed a correlation between an increase in the amount of TAU protein and wound response of apples.

An induction of the expression of *β-glucosidase13* and *β-glucosidase40* genes in response to inoculation by *P. digitatum* and *P. expansum*, respectively, was also observed (Vilanova et al. 2014b). Sánchez-Torres and González-Candelas (2003) found that the glycosidase gene was specifically expressed in apple tissue infected by *P. expansum*. The first evidence of the expression of the *β-glucosidase* gene in plants was reported by Simmons et al. (2001), that correlated upregulation of the *rhm1* corn mutant to resistance to *Bipolaris maydis*.

Gene Expression of Genes Involved in the Phenylpropanoid Pathway

Because of the involvement of lignin in defence processes of oranges and apples against biotic and abiotic stresses, a gene expression study of several phenylpropanoid pathway-related genes was conducted for apples (Vilanova et al. 2014b) and oranges (Vilanova et al. 2013).

Apples

Microarray study revealed changes in the expression of genes involved in phenylpropanoid pathway, as the phenylalanine ammonium liase (*PAL1* and *PAL2*), caffeic acid O-methyltransferase (*COMT*) and cinnamoyl-CoA reductase (*CCR*) genes, in apple inoculated with *P. digitatum* (non-host pathogen). Transcription levels of some genes (*PAL1*, *PAL2*, *COMT2* and *POX64*) were evaluated in apples at immature and commercial harvest, inoculated with *P. expansum*, *P. digitatum* and water (control) at 8, 24 and 48 h post inoculation (hpi) using the quantitative PCR (qPCR) technique.

At 24 hpi, higher transcript abundance of *PAL1*, *PAL2* and *POX64* genes was observed for both apples inoculated by the non-host pathogen (*P. digitatum*) than other treatments. In contrast, at 48 hpi the highest increase in gene expression was found in *PAL1*, *PAL2*, *COMT2* and *POX64* in apples inoculated with *P. expansum*. Ballester et al. (2013) found that some genes related to the phenylpropanoid pathway showed their maximum expression at 48 hpi and decreased at 72 hpi. These variances in the expression along the time may be due to differences in inoculum concentrations. Furthermore, these results are in agreement with that observed in histochemical studies: in a short-term response both fungi showed a positive reaction to lignin, while in a long-term response only a positive reaction was obtained for the non-host pathogen, because the other samples were completely rotten (Vilanova et al. 2012a). Overall results support the hypothesis that apple attempt to defend against both pathogens, although host pathogen is finally able to overcome apple defences and develop infection.

Regarding apple maturity stages, microarray study indicated that *PAL1*, *PAL2* and *POX64* genes showed a greater abundance in commercial than in immature apples. Both PAL and POX belong to a family of enzymes that are involved in various biochemical functions in plants as growth, development and senescence. High PAL activity has also been associated with the maturity stage of apples (Wang et al. 2000) and strawberries (Cheng and Breen 1991; Villarreal et al. 2010) due to the accumulation of anthocyanins and other phenolic compounds. An increase in POX activity during ripening has been observed in apples (Torres et al. 2003) and plums (Singh et al. 2012). Therefore, the observed increase in the expression of *PAL1*, *PAL2* and *POX64* genes may be more related to fruit ripening rather than defence responses.

Oranges

In previous studies, Ballester et al. (2011, 2013) analysed the gene expression of several genes involved in the phenylpropanoid pathway and the modifications in phenolic compounds associated with citrus resistance against *P. digitatum*. In Vilanova et al. (2013), some of these genes were analysed in immature oranges infected by *P. digitatum* but also by *P. expansum*, because the response of citrus fruit to non-host pathogen remained unexplored.

In oranges, the expression of *PAL1*, *COMT1*, *POX1*, cinnamyl alcohol dehydrogenase (*CAD*) and sinapyl alcohol dehydrogenase (*SAD*) genes was analyzed after inoculation with *P. digitatum*, *P. expansum* and water (control), 24 and 48 hpi using the semi-quantitative PCR technique (Vilanova et al. 2013). Gene expression analysis showed different behaviour depending on the time after inoculation. At 24 hpi, oranges inoculated with the compatible pathogen (*P. digitatum*) showed a higher expression of the *PAL1* and *COMT1* genes compared to oranges inoculated with

P. expansum and control, although the *POX1* gene showed greater expression in oranges inoculated with both moulds than control. At 48 h, however, oranges inoculated with *P. digitatum* displayed an important decrease in expression of *PAL1*, *COMT1*, *POX1* and *SAD* genes, and oranges inoculated with *P. expansum* and control the gene expression remained constant. This decrease in expression seems to indicate that the compatible pathogen is able to suppress the gene expression related to the phenylpropanoid pathway and therefore start the infection.

Concluding Remarks

Our studies suggest that maturity stage is an important factor to be considered on pathological studies. Therefore, it would be interesting to continue exploring the relationship between maturity factors and fruit susceptibility in response to pathogens. Besides, the increase of resistance or susceptibility of apples and oranges to host and non-host pathogens depend on it and may vary based on the climacteric or non-climacteric respiration pattern of fruit.

Depending on the combination of certain conditions as maturity stage, inoculum concentration and storage temperature, the non-host pathogen become compatible pathogen in these fruit –pathogen interactions. Furthermore, wounded fruit stored at cold temperatures are, in general, more susceptible to pathogen attack. Lignification is a key factor in fruit defence to both biotic and abiotic stresses together with other multiple reactions. This process was apparently more important in immature than in commercial and over-mature fruit.

Finally, the global study conducted on apples give us an overall picture of metabolic pathways to may activate in response to host and non-host pathogens to continue to work in this research line with the aim of unravelling the fruit defence mechanisms.

Overall knowledge obtained over the time-course of our studies can help us to establish the basis towards the improvement of postharvest control strategies in apples and oranges.

References

Baert K, Devlieghere F, Flyps H, Oosterlinck M, Ahmed MM, Rajković A, Verlinden B, Nicolaï B, Debevere J, De Meulanaer B (2007a) Influence of storage conditions of apples on growth and patulin production by *Penicillium expansum*. Int J Food Microbiol 119:170–181

Baert K, Valero A, De Meulenaer B, Samapundo S, Ahmed MM, Bo L, Debevere J, Devlieghere F (2007b) Modeling the effect of temperature on the growth rate and lag phase of *Penicillium expansum* in apples. Int J Food Microbiol 118:139–150

Ballester AR, Lafuente MT, González-Candelas L (2006) Spatial study of antioxidant enzymes, peroxidase and phenylalanine ammonia-lyase in the citrus fruit-*Penicillium digitatum* interaction. Postharvest Biol Tec 39:115–124

Ballester AR, Izquierdo A, Lafuente MT, González-Candelas L (2010) Biochemical and molecular characterization of induced resistance against *Penicillium digitatum* in citrus fruit. Postharvest Biol Tec 56:31–38

Ballester AR, Lafuente MT, Forment J, Gadea J, De Vos RC, Bovy AG, González-Candelas L (2011) Transcriptomic profiling of citrus fruit peel tissues reveals fundamental effects of phenylpropanoids and ethylene on induced resistance. Mol Plant Pathol 12:879–897

Ballester AR, Lafuente MT, González-Candelas L (2013) Citrus phenylpropanoids and defence against pathogens. Part II: gene expression and metabolite accumulation in the response of fruits to *Penicillium digitatum* infection. Food Chem 136:285–291

Baudoin A, Eckert JW (1985) Development of resistance against *Geotrichum candidum* in lemon peel injuries. Phytopathology 75:174–179

Bocsanczy A, Phillips JG, Dardick CD, Korban SS, Bassett CL, Wisniewski ME, Norelli JL (2009) Analysis of apple (*Malus*) responses to bacterial pathogens using an oligo microarray. Phytopathology 99:S14–S14

Brown GE (1989) Host defenses at the wound site on harvested crops. Phytopathology 79:1381–1384

Brown GE, Barmore CR (1983) Resistance of healed citrus exocarp to penetration by *Penicillium digitatum*. Phytopathology 73:691–694

Brown GE, Ismai MA, Barmore CR (1978) Lignification of injuries to citrus fruit and susceptibility to green mold. Proc Florida State Hortic Soc 91:124–126

Buron-Moles G, Lopez-Perez M, González-Candelas L, Viñas I, Teixidó N, Usall J, Torres R (2012) Use of GFP-tagged strains of *Penicillium digitatum* and *Penicillium expansum* to study host-pathogen interactions in oranges and apples. Int J Food Microbiol 160:162–170

Buron-Moles G, Torres R, Amoako-Andoh F, Viñas I, Teixidó N, Usall J, Keulemans W, Davey MW (2014) Analysis of changes in protein abundance after wounding in 'golden delicious' apples. Postharvest Biol Tec 87:51–60

Cantu D, Vicente AR, Greve LC, Dewey FM, Bennett AB, Labavitch JM, Powell AL (2008a) The intersection between cell wall disassembly, ripening, and fruit susceptibility to *Botrytis cinerea*. Proc Natl Acad Sci USA 105:859–864

Cantu D, Vicente AR, Labavitch JM, Bennett AB, Powell AL (2008b) Strangers in the matrix: plant cell walls and pathogen susceptibility. Trends Plant Sci 13:610–617

Cheng GW, Breen PJ (1991) Activity of phenylalanine ammonia-lyase (PAL) and concentrations of anthocyanins and phenolics in developing strawberry fruit. J Am Soc Hortic Sci 116:865–869

Chiu CM, You BJ, Chou CM, Yu PL, Yu FY, Pan SM, Bostock RM, Chung KR, Lee MH (2013) Redox status-mediated regulation of gene expression and virulence in the brown rot pathogen *Monilinia fructicola*. Plant Pathol 62:809–819

De Gara L, De Pinto MC, Tommasi F (2003) The antioxidant systems vis-a-vis reactive oxygen species during plant-pathogen interaction. Plant Physiol Biochem 41:863–870

Droby S, Chalutz E, Horev B, Cohen L, Gaba V, Wilson CL, Wisniewski M (1993) Factors affecting UV-induced resistance in grapefruit against the green mould decay caused by *Penicillium digitatum*. Plant Pathol 42:418–424

Droby S, Porat R, Cohen L, Weiss B, Shapiro B, Philosoph-Hadas S, Meir S (1999) Suppressing green mold decay in grapefruit with postharvest jasmonate application. J Am Soc Hortic Sci 124:184–188

European Apple and Pear Crop Forecast. Prognosfruit (August 2018). www.Prognosfruit.eu

Food and Agriculture Organization of the United Nations (FAO) (2017) Citrus fruit, fresh and processed statistical bulletin 2016 – http://www.fao.org/3/a-i8092e.pdf. Accessed 15 Apr 2020

Govrin EM, Levine A (2000) The hypersensitive response facilitates plant infection by the necrotrophic pathogen *Botrytis cinerea*. Curr Biol 10:51–757

Hatfield RD, Grabber J, Ralph J, Brei K (1999) Using the acetyl bromide assay to determine lignin concentrations in herbaceous plants: some cautionary notes. J Agric Food Chem 47:628–632

Hershkovitz V, Ben-Dayan C, Raphael G, Pasmanik-Chor M, Liu J, Belausov E, Aly R, Wisniewski M, Droby S (2012) Global changes in gene expression of grapefruit peel tissue in response to the yeast biocontrol agent *Metschnikowia fructicola*. Mol Plant Pathol 13:338–349
Ismail MA, Brown GE (1975) Phenolic content during healing of 'Valencia' orange peel under high humidity. J Am Soc Hortic Sci 100:249–251
Kavanagh J, Wood R (1967) Role of wounds in infection of oranges by *Penicillium digitatum* Sacc. Ann App Biol 60:375–383
Kim SG, Kim YH, Kim HT (2008) Effect of delayed inoculation after wounding on the development of anthracnose disease caused by *Colletotrichum acutatum* on chilli pepper fruit. Plant Pathol J 24:392–399
Lai S, Lai A, Stange Jr, McCollum TG, Schirra M (2003) Characterization of the wound-induced material in *Citrus paradisi* fruit peel by carbon-13 CP-MAS solid state NMR spectroscopy. Phytochemistry 63:177–183
Lakshiminarayana S, Sommer NF, Polito V, Fortlage RJ (1987) Development of resistance to infection by *Botrytis cinerea* and *Penicillium expansum* in wounds of mature apple fruits. Phytopathology 77:1674–1678
Lamb C, Dixon RA (1997) The oxidative burst in plant disease resistance. Annu Rev Plant Physiol Plant Mol Biol 48:251–275
Lauxmann MA, Brun B, Borsani J, Bustamante CA, Budde CO, Lara MV, Drincovich MF (2012) Transcriptomic profiling during the post-harvest of heat-treated dixiland prunus persica fruits: common and distinct response to heat and cold. PLoS One 7:e51052
Lee YP, Yu GH, Seo YS, Han SE, Choi Y-O, Kim D, Mok I-G, Kim WT, Sung S-K (2007) Microarray analysis of apple gene expression engaged in early fruit development. Plant Cell Rep 26:917–926
Levine A, Tenhaken R, Dixon R, Lamb C (1994) H_2O_2 from the oxidative burst orchestrates the plant hypersensitive disease resistance response. Cell 79:583–593
Louw JP, Korsten L (2014) Pathogenic *Penicillium* spp. on apple and pear. Plant Dis 98:590–598
Low PS, Merida JR (1996) The oxidative burst in plant defense: function and signal transduction. Physiol Plant 96:533–542
Lu HG, Higgins VJ (1999) The effect of hydrogen peroxide on the viability of tomato cells and of the fungal pathogen *Cladosporium fulvum*. Physiol Mol Plant Pathol 54:131–143
Macarisin D, Cohen L, Eick A, Rafael G, Belausov E, Wisniewski M, Droby S (2007) *Penicillium digitatum* suppresses production of hydrogen peroxide in host tissue infection of citrus fruit. Phytopathology 97:1491–1500
Mayer AM, Staples RC, Gil-ad NL (2001) Mechanisms of survival of necrotrophic fungal plant pathogens in hosts expressing the hypersensitive response. Phytochemistry 58:33–41
Mittler R, Lam E, Shulaev V, Cohen M (1999) Signals controlling the expression of cytosolic ascorbate peroxidase during pathogen-induced programmed cell death in tobacco. Plant Mol Biol 39:1025–1035
Morales H, Sanchis V, Coromines J, Ramos AJ, Marín S (2008) Inoculum size and intraspecific interactions affects *Penicillium expansum* growth and patulin accumulation in apples. Food Microbiol 25:378–385
Mulas M, Lafuente MT, Zacarías L (1996) Lignin and gum deposition in wounded 'Oroval' clementines as affected by chilling and peel water content. Postharvest Biol Tec 7:243–251
Mysore KS, Ryu CM (2004) Nonhost resistance: how much do we know? Trends Plant Sci 9:97–104
Neri F, Donati I, Veronesi F, Mazzoni D, Mari M (2010) Evaluation of *Penicillium expansum* isolates for aggressiveness, growth and patulin accumulation in usual and less common fruit hosts. Int J Food Microbiol 143:109–117
Pavez L, Hödar C, Olivares F, González M, Cambiazo V (2013) Effects of postharvest treatments on gene expression in *Prunus persica* fruit: normal and altered ripening. Postharvest Biol Tec 75:125–134

Pavoncello D, Lurie S, Droby S, Porat R (2001) A hot water treatment induces resistance to *Penicillium digitatum* and promotes the accumulation of heat shock and pathogenesis-related proteins in grapefruit flavedo. Physiol Plant 111:17–22

Plaza P, Usall J, Teixidó N, Viñas I (2003) Effect of water activity and temperature on germination and growth of *Penicillium digitatum, P. italicum* and *Geotrichum candidum*. J Appl Microbiol 94:549–554

Quaglia M, Ederli L, Pasqualini S, Zazzerini A (2011) Biological control agents and chemical inducers of resistance for postharvest control of *Penicillium expansum* Link. on apple fruit. Postharvest Biol Tec 59:307–315

Rolke Y, Liu SJ, Quidde T, Williamson B, Schouten A, Weltring KM, Siewers V, Tenberge KB, Tudzynski B, Tudzynski P (2004) Functional analysis of H_2O_2-generating systems in *Botrytis cinerea*: the major Cu-Zn-superoxide dismutase (BCSOD1) contributes to virulence on French bean, whereas a glucose oxidase (BCGOD1) is dispensable. Mol Plant Pathol 5:17–27

Romero D, Rivera ME, Cazorla FM, Codina JC, Fernández-Ortuño D, Torés JA, Pérez-García A, de Vicente A (2008) Comparative histochemical analyses of oxidative burst and cell wall reinforcement in compatible and incompatible melon-powdery mildew (*Podosphaera fusca*) interactions. J Plant Physiol 165:1895–1905

Sánchez-Torres P, González-Candelas L (2003) Isolation and characterization of genes differentially expressed during the interaction between apple fruit and *Penicillium expansum*. Mol Plant Pathol 4:447–457

Sarowar S, Zhao Y, Soria-Guerra RE, Ali S, Zheng D, Wang D, Korban SS (2011) Expression profiles of differentially regulated genes during the early stages of apple flower infection with *Erwinia amylovora*. J Exp Bot 62:4851–4861

Shao X, Tu K, Tu S, Zhao Y (2010) Effects of heat treatment on wound healing in Gala and Red Fuji apple fruits. J Agric Food Chem 58:4303–4309

Silva MC, Nicole M, Guerra-Guimarães L, Rodrigues CJ Jr (2002) Hypersensitive cell death and post-haustorial defence responses arrest the orange rust (*Hemileia vastatrix*) growth in resistant coffee leaves. Physiol Mol Plant Pathol 60:169–183

Simmons CR, Grant S, Altier DJ, Dowd PF, Crasta O, Folkerts O, Yalpani N (2001) Maize rhm1 resistance to *Bipolaris maydis* is associated with few differences in pathogenesis-related proteins and global mRNA profiles. Mol Plant-Microbe Interact 14:947–954

Singh SP, Singh Z, Swinny EE (2012) Climacteric level during fruit ripening influences lipid peroxidation and enzymatic and non-enzymatic antioxidative systems in Japanese plums (*Prunus salicina* Lindell). Postharvest Biol Tec 65:22–32

Soria-Guerra RE, Rosales-Mendoza S, Gasic K, Wisniewski ME, Band M, Korban SS (2011) Gene expression is highly regulated in early developing fruit of apple. Plant Mol Biol Rep 29:885–897

Spotts RA, Sanderson PG, Lennox CL, Sugar D, Cervantes LA (1998) Wounding, wound healing and staining of mature pear fruit. Postharvest Biol Tec 13:27–36

Stange RR, Midland SL, Sims JJ, McCollum TG (2002) Differential effects of citrus peel extracts on growth of *Penicillium digitatum, P. italicum*, and *P. expansum*. Physiol Mol Plant Pathol 61:303–311

Su J, Tu K, Cheng L, Tu S, Wang M, Xu H, Zhan G (2011) Wound-induced H_2O_2 and resistance to *Botrytis cinerea* decline with the ripening of apple fruit. Postharvest Biol Tec 62:64–70

Tian S, Torres R, Ballester AR, Li B, Vilanova L, González-Candelas L (2016) Molecular aspects in pathogen-fruit interactions: virulence and resistance. Postharvest Biol Tec 122:11–21

Torres R, Valentines MC, Usall J, Viñas I, Larrigaudiere C (2003) Possible involvement of hydrogen peroxide in the development of resistance mechanisms in 'golden delicious' apple fruit. Postharvest Biol Tec 27:235–242

Torres MA, Jones JDG, Dangl JL (2006) Reactive oxygen species signaling in response to pathogens. Plant Physiol 141:373–378

Valentines MC, Vilaplana R, Torres R, Usall J, Larrigaudiere C (2005) Specific roles of enzymatic browning and lignification in apple disease resistance. Postharvest Biol Tec 36:227–234 van Loon LC, van Strien EA (1999) The families of pathogenesis-related proteins, their activities, and comparative analysis of PR-1 type proteins. Physiol Mol Plant Pathol 55:85–97

van Loon LC, Rep M, Pieterse CMJ (2006) Significance of inducible defense-related proteins in infected plants. Annu Rev Phytopathol 44:135–162

Vierling E (1991) The roles of heat-shock proteins in plants. Ann Rev Plant Physioly and Plant Mol Biol 42:579–620

Vilanova l, Teixidó N, Torres R, Usall J, Viñas I (2012a) The infection capacity of *P. expansum* and *P. digitatum* on apples and histochemical analysis of host response. Int J Food Microbiol 157:360–367

Vilanova l, Viñas I, Torres R, Usall J, Jauset AM, Teixidó N (2012b) Infection capacities in the orange-pathogen relationship: compatible (*Penicillium digitatum*) and incompatible (*Penicillium expansum*) interactions. Food Microbiol 29:56–66

Vilanova L, Torres R, Viñas I, González-Candelas L, Usall J, Fiori S, Solsona C, Teixidó N (2013) Wound response in oranges as a resistance mechanism against *Penicillium digitatum* (pathogen) and *P. expansum* (non-host pathogen). Postharvest Biol Tech 78:113–122

Vilanova L, Viñas I, Torres R, Usall J, Buron-Moles G, Teixidó N (2014a) Increasing maturity reduces wound response and lignification process against *Penicillium expansum* (pathogen) and *Penicillium digitatum* (non-host pathogen) infection in apples. Postharvest Biol Tech 88:54–60

Vilanova L, Wisniewski M, Norelli J, Viñas I, Torres R, Usall J, Phillips J, Droby S, Teixidó N (2014b) Transcriptomic profiling of apple in response to inoculation with a pathogen (*Penicillium expansum*) and a non-pathogen (*Penicillium digitatum*). Plant Mol Biol Rep 32:566–583

Villarreal NM, Bustamante CA, Civello PM, Martínez GA (2010) Effect of ethylene and 1-MCP treatments on strawberry fruit ripening. J Sci Food Agri 90:683–689

Wang HQ, Arakawa O, Motomura Y (2000) Influence of maturity and bagging on the relationship between anthocyanin accumulation and phenylalanine ammonia-lyase (PAL) activity in 'Jonathan' apples. Postharvest Biol Tec 19:123–128

Yun Z, Gao HJ, Liu P, Liu S, Luo T, Jin S, Xu Q, Xu J, Cheng Y, Deng X (2013) Comparative proteomic and metabolomic profiling of citrus fruit with enhancement of disease resistance by postharvest heat treatment. BMC Plant Biol 16:13–44

Zhu H, Dardick CD, Beers EP, Callanhan AM, Xia R, Yuan R (2011) Transcriptomics of shading-induced and NAA-induced abscission in apple (*Malus domestica*) reveals a shared pathway involving reduced photosynthesis, alterations in carbohydrate transport and signaling and hormone crosstalk. BMC Plant Biol 11:138–158

Chapter 3
Uncovering the NLR Family of Disease Resistance Genes in Cultivated Sweetpotato and Wild Relatives

Camilo H. Parada Rojas and Lina M. Quesada-Ocampo

Abstract Sweetpotato (*Ipomoea batatas*) is an important staple crop cultivated on every continent except Antarctica. Awareness of sweetpotato nutritional benefits have led to an increase in consumption in the United States (US) and Europe. Despite continued implementation of disease management strategies, the fungal pathogen *Ceratocystis fimbriata* persists as a significant threat to the sweetpotato industry in the US. The presence of few breeding programs, limited knowledge of resistance, and the hexaploid nature of sweetpotato pose challenges to develop resistant lines. Plants possess an innate immune system with disease resistance (R) genes encoding for proteins that recognize pathogen effectors during infection. An important class of resistance gene contains nucleotide-binding and leucine-rich repeat domains, NLRs. Here, we review state of the art knowledge of sweetpotato NLRs as well as novel methods to predict NLRs in plant genomes. Despite the availability of chemical control options, fungicide use is becoming limited due to changes in regulation and the evolution of pathogen resistance to chemicals. Deployment of host resistance represents a desirable tool to decrease crop losses due to plant pathogens. NLR gene annotations provide significant insight into the resistome and a first step towards identifying genes effective for control of major plant pathogens of sweetpotato.

Keywords *Ipomoea batatas* · NLRs · Next-generation sequencing · Pathogen resistance · Comparative genomics

C. H. Parada Rojas · L. M. Quesada-Ocampo (✉)
Department of Entomology and Plant Pathology, North Carolina State University, Raleigh, NC, USA
e-mail: lmquesad@ncsu.edu

D. Spadaro et al. (eds.), *Postharvest Pathology*, Plant Pathology in the 21st Century 11, https://doi.org/10.1007/978-3-030-56530-5_3

Introduction

Importance of Sweetpotato

Sweetpotato (*Ipomoea batatas* (L.) Lam) is native to Latin America and the fifteenth most important food crop worldwide based on its production in tons. Sweetpotatoes are cultivated on about 9.2 million hectares, of which 53% are grown in low income food deficit countries (FAOSTAT 2017b). Today, China is the largest producer of sweetpotato, accounting for 36% of the total area of sweetpotato cultivated in the world. In 2016, China produced about 72 million tons with the vast majority of its output going to animal consumption (Bradshaw 2010; FAOSTAT 2017a; Sheikha and Ray 2017). The use of sweetpotato for animal feed plays an important role not only in China but in the rural economies of Philippines, India, Korea, Taiwan, Papua New Guinea and, to a lesser extent, in Latin America and Africa (Loebenstein 2009). The United States (US) currently rank tenth in terms of global annual production (FAOSTAT 2017b). Value-added products such as sweetpotato fries and chips, as well as awareness of nutritional benefits have led to an increase in sweetpotato consumption in the US and Europe (Scruggs and Quesada-Ocampo 2016; Sheikha and Ray 2017). Sweetpotatoes are considered a specialty crop in the US with a total annual production value of 654 million dollars in 2018 (USDA-NASS 2018). In the same year, US sweetpotato exports peaked at 336,932 tons with a value of 192 million dollars (USDA-ERS 2018). In 2016, the US ranked first trading 44% of the global market of sweetpotatoes for human consumption (FAOSTAT 2016).

Sweetpotato is a valuable source of carbohydrates, fibers, iron, vitamin A, C, and some of the B-complex vitamins (USDA 2017). Orange-fleshed sweetpotato varieties harbor ß-carotene, the precursor of vitamin A, which makes them desirable as a food source in sub-Saharan Africa; where vitamin A deficiencies contribute to high rates of blindness, disease, and premature death in children and pregnant women (Sheikha and Ray 2017). Researchers reported that purple fleshed sweetpotato varieties contain anthocyanins, a powerful antioxidant that could scavenge free radicals, inhibit cancer cell growth, and attenuate inflammatory response (Wang and Stoner 2008; Karna et al. 2011; Saigusa et al. 2005). Historically, sweetpotato has been invaluable during periods of famine substituting other staple crops. Sweetpotato replaced demolished rice fields after typhoon flooding in Japan, saved a great number of lives in China during a famine in 1954, and supplanted cassava crops in Uganda when a virus decimated them in the 1990s (CIP 2010; Yan et al. 2014; Zhang et al. 2009).

Origin and History of Sweetpotato

Before the first contact with European conquistadors or Polynesian voyagers, the Arawak Indians lived in village communes and had developed agriculture of corn, yams, cassava, and sweetpotato (Zinn 2015). Archaeological records dated to

8000 years ago suggest that the *I. batatas* was domesticated in the Peruvian highlands and in northern South America (Lebot 2010; Loebenstein and Thottappilly 2009). Austin (1987) offered taxonomic evidence as to how *I. batatas* natural hybridization occurred. He proposed that *I. trifida* and *I. triloba* generated the wild ancestors of *I. batatas* somewhere between the Yucatan Peninsula of Mexico and the mouth of the Orinoco River in Venezuela. More recently, the analysis of a representative collection of sweetpotato landraces (329 accessions) based on a set of nuclear and chloroplast microsatellite markers revealed the existence of two geographically restricted gene pools. One corresponding to accessions from southern Colombia, Ecuador, and Peru and the other corresponding to accessions from Mexico to northern Colombia and Venezuela (Roullier et al. 2011). Between 3000 and 31,100 years ago, Polynesian voyagers visited and collected sweetpotatoes from the western coast of South America, marking the first expansion of sweetpotato from the Americas into Oceania (Clarke et al. 2006). After the European invasion, sweetpotato disembarked in Europe and subsequently explorers dispersed the crop to Africa and India (Loebenstein 2009). Around one hundred years after the European invasion, Spanish galleons and Portuguese traders introduced the Mesoamerican sweetpotato lineage and the Caribbean sweetpotato lineage respectively into the Philippines and Indonesia. This last dispersal event represents the second and third expansion of sweetpotatoes from the Americas into Oceania (Roullier et al. 2013). Secondary dispersal events in the late sixteenth century introduced sweetpotato from the Philippines into China (Roullier et al. 2013; O'Brien 1972). Several accounts describe the arrival or presence of sweetpotato in North America around the 1500s for wild sweetpotatoes and 1700s for cultivated sweetpotatoes (O'Brien 1972).

Sweetpotato Production

At the morphological level, sweetpotatoes are different from potato, *Solanum tuberosum* L., which belongs to the Solanaceae family and produce swollen subterranean stems known as tubers (Villordon et al. 2014). Sweetpotatoes belong to the Convolvulaceae family and are a true root (Smith et al. 2009; Wu et al. 2018b). A sweetpotato plant can yield between 2 and 15 storage roots and storage formation is associated with thickening of adventitious roots (Villordon et al. 2014; Yencho et al. 2008). In the US, sweetpotato production spans across wide geographical regions and soil types (USDA-NASS 2018). North Carolina, Mississippi, California, Louisiana, Florida, and Arkansas represent the top sweetpotato producing states accounting for more than 90% of the sweetpotato produced in the US (USDA-NASS 2018). Smith et al. (2009) defined sweetpotato as a "high risk" crop due to the potential for weather-related failures, and the aesthetic qualities of the crop that can be affected by insect feeding and disease incidence. In fact, every year the United States Department of Agriculture issues sweetpotato crop insurance coverage to alleviate crop related losses (USDA 2016). Sweetpotatoes are vegetatively propagated using cuttings obtained from root sprouts also known as slips (Loebenstein and Thottappilly 2009). Sweetpotato production in the US comprises three distinct operational units: transplant preparation, field production, and packing. A sweetpotato network of seed producing growers every year buries storage roots on seedbeds to produce slips. During May through July, slips are cut approximately 25 – 35 cm long above the soil surface. By cutting slips, growers reduce the potential for moving root diseases from the seedbeds into the production fields (Holmes and Kemble 2010). The crop growing period varies from 110 to 130 days, depending on cultivar and transplant spacing (Boudreaux 2005). Most sweetpotato growers transplant their slips using tobacco mechanical transplanters. The transplanters ensure that the slip is place underground at a depth of 10–15 cm with at least 2–3 nodes underground to maximize yield potential (Smith et al. 2009). Cultivated conditions have been reviewed before (Lebot 2010; O'Brien 1972; Loebenstein and Thottappilly 2009; Zhang et al. 2009; Smith et al. 2009). For most part these reviews provide key cultivation literature and can be a good reference. We would like to emphasize that the majority of sweetpotato producers in the US utilize virus-tested tissue culture technology and supplement their conventional slips produced on-farm with certified virus-tested slips (Kokkinos and Clark 2006). Virus testing has become an integral part of slip production programs led by North Carolina State University, Louisiana State University, among other institutions. Implementation of this technology results in significant yield increase, reduction of somatic mutations, and lower incidence of pathogens (Clark et al. 2012).

Sweetpotato field harvest usually begins in early August and extends until November or later if necessary (Holmes and Kemble 2010). Most sweetpotato farmers combine mechanical harvesting with hand harvest using potato chain harvesters or farmer workers respectively. Immediately after harvest, roots are deposited into wooden or plastic bins (~ 454 kg) and subsequently moved into storage facilities to undergo curing (Smith et al. 2009). The curing process involves holding roots at a temperature of 29 °C and 85% relative humidity with proper ventilation for 5–7 days immediately after harvest. Curing aids in wound healing by enhancing suberization, which reduces water loss and disease progression (Edmunds et al. 2008). Cured roots are stored in climate controlled facilities for long term at 13 °C and 85% relative humidity which gives producers and brokers leverage in moving and marketing the crop throughout the year (Scruggs and Quesada-Ocampo 2016; Smith et al. 2009). Every year, sweetpotato production is limited by different abiotic and biotic factors, among which plant pathogens constitute a significant threat to the industry worldwide. A few of the most notable sweetpotato plant pathogens include *Fusarium solani* (Fusarium root rot), *Monilochaetes infuscans* (Sweetpotato Scurf), *Streptomyces ipomoeae* (Streptomyces soil rot), *Macrophomina phaseolina* (Charcoal rot), *Rhizopus stolonifer* (Rhizopus soft rot), *Meloidogyne incognita* (root-knot nematode) and the recently re-emerging soilborne pathogen *Ceratocystis fimbriata*, the causal agent of black rot (Clark et al. 2013; Scruggs et al. 2017).

The Sweetpotato Genome

Thanks to the next-generation sequencing revolution, plant genomics has transformed drastically in the past 20 years, resulting in genome sequencing projects in more than 100 plant species (Michael and VanBuren 2015). The high throughput, low cost, and longer reads of next-generation sequencing technologies coupled with advances in bioinformatics made it possible to sequence non-model plant species revealing important clues to plant genome architecture and plant evolution (Michael and Jackson 2013). Despite these developments, most polyploid plant genomes have not yet been sequenced. The difficulty of distinguishing chromosome duplication events among genome sequences limits efforts to sequence whole genomes of polyploid species (Isobe et al. 2017). Consequently, genome survey sequencing of ancestor diploid species is often used to overcome this limitation by reducing the complexity of genome structure (Bertioli et al. 2016; Ling et al. 2013).

To date, the sweetpotato whole genome sequence has not been published. The main reason is its complex structure with a genome size of about 4.4 Gb (hexaploid, 6X) and its high degree of heterozygosity (Isobe et al. 2017; Ozias-Akins and Jarret 1994). Several *Ipomoea* species have been hypothesized as candidate ancestor species of *I. batatas* including *I. triloba*, *I. tabascana*, and *I. trifida*. However, *I. trifida* is suggested as the most probable diploid wild relative of *I. batatas* based on cytogenetic, molecular genetics, and cross-compatibility (Wu et al. 2018b; Hirakawa et al. 2015). As a consequence, two genome surveys of *I. trifida* have been reported. First, Hirakawa et al. (2015) used the Illumina HiSeq platform to produce the first reported *de novo* whole-genome sequence with two lines of *I. trifida*, a selfed line Mx23Hm and the highly heterozygous line 0431-1. The authors identified a total of 62,407 and 109,449 putative genes in the genomes of Mx23Hm and 0431-1, respectively. Recently, a collaborative effort delivered high-quality *I. trifida* and *I. triloba* genome sequences and developed a searchable genome sequence-based marker platform for sweetpotato improvement (Wu et al. 2018b). This project involved seven organizations including the International Potato Center (Peru, Uganda), two US universities (NC State University and Michigan State University), a US institute (Boyce Thompson Institute), an Australian institute (Institute for Molecular Bioscience), and a Chinese research organization (Food Crops Institute).

Currently, research groups worldwide are taking part in the challenge to produce the first whole genome assembly for *I. batatas* using slightly different approaches. In September 2017, using a combined genome assembly and haplotyping method, Yang et al. (2017) published a half haplotype-resolved genome for a newly bred carotenoid-rich cultivar of *I. batatas*, Taizhong6. This represents the first monoploid genome of sweetpotato. The final assembly of ~836 Mb and scaffold N50 of ~210 kb was generated entirely based on Illumina sequencing. Likewise, a sweetpotato genome consortium consisting of research groups from China, Korea, and Japan are working on a genome sequence of the sweetpotato cultivar, Xushu 18. Their goal is to perform *de novo* whole genome sequencing, transcript analysis, and linkage map construction by using Illumina and PacBio platforms (Isobe et al. 2017).

A Case Study: *Ceratocystis fimbriata*, The Causal Agent of Black Rot of Sweetpotato

Aiming to shed some light upon the nature of sweetpotato diseases affecting New Jersey farmers, Halsted (1890) published the first description of the ascomycete soilborne fungus *Ceratocystis fimbriata* E & Halsted in the Bulletin No. 76 of the New Jersey Agricultural College Experiment Station. Halsted named the disease black rot because of the presence of dark-brown patches with irregular shape in the roots (Fig. 3.1). He described the main concern with black rot as "the root lesions are usually evident at digging-time, but are small enough that they find their way into the market and further progression of the disease falls upon the middlemen and consumers" (Halsted 1890). *C. fimbriata* causes severe infections on a wide range of plants worldwide, including cacao, coffee, coconut, sunn hemp, sycamore, poplar, almond, oak, taro, mango and fig, among others. In sweetpotato, under favorable conditions, the infection results in black, sunken cankers on sprouts or the

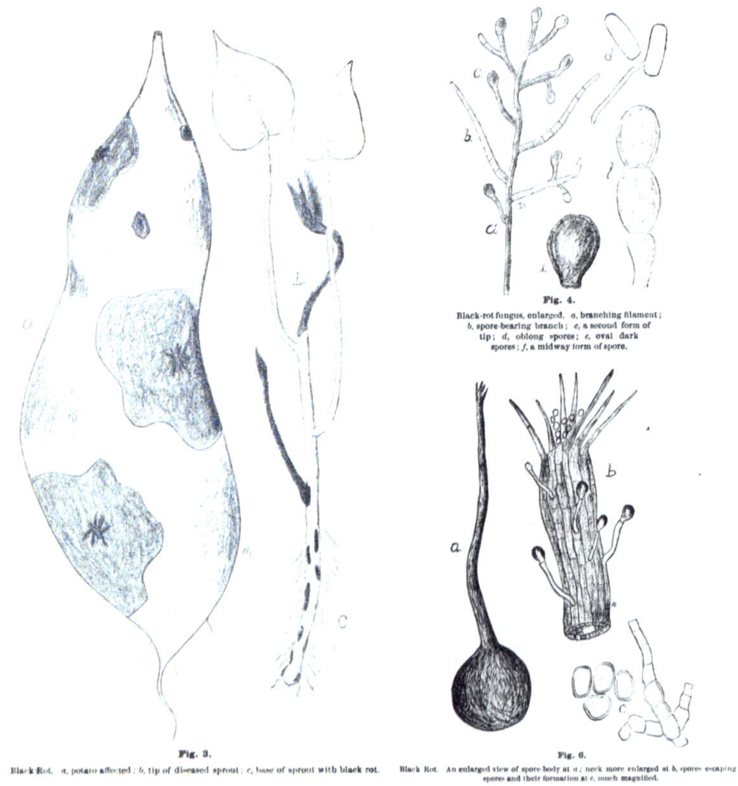

Fig. 3.1 Byron D. Halsted morphological description of *Ceratocystis fimbriata*. (Taken from the Bulletin No. 76 at New Jersey Agriculture College Experiment Station)

Fig. 3.2 Typical black rot symptoms caused by *Ceratocystis fimbriata*. (**a**) Dark concentric dry lesion on sweetpotato storage root (**b**) sunken cankers in slips or the developing roots

developing roots, leading to stunting, wilting, chlorosis, and eventually death (Fig. 3.2). Storage conditions favor the development of black, firm lesions in the root surface often producing a sweet, fruity odor (Clark et al. 2013; Scruggs et al. 2017). Under suitable environmental conditions, *C. fimbriata* colonizes wounds and slowly develops lesions that are typically restricted by the vascular ring. The center of the lesions often produces perithecia and sticky ascospores that aid in dissemination. These represent active propagules that contaminate equipment and lead to severe losses (Scruggs et al. 2017). The fungus has been reported to survive in the soil for extended periods, however little is known about the soil-borne phase of the *C. fimbriata* disease cycle (Engelbrecht 2004).

In 2015, black rot reemerged as a significant threat to sweetpotato production in North Carolina (NC) and was observed postharvest in numerous other locations throughout the US, including Virginia, Georgia, Mississippi, Alabama, Louisiana, Arkansas, Missouri, and California, as well as in Ontario, Canada. Heavy yield losses were reported and confirmed in the Plant Disease and Insect Clinic at NC State University from across NC. The reemergence of *C. fimbriata* prompted research to improve control options and determine the cause of the recent outbreaks. The following sections are an overview of the biology, ecology, genetics, and management of black rot in sweetpotato.

Management of Black Rot in Sweetpotato

Historically, black rot was effectively controlled by cutting transplants above the soil line, using thiabendazole based fungicides on seed roots, reducing wounds during harvesting, and sanitizing packing equipment (Clark 1992). The implementation of integrated control strategies led to a drastic reduction in disease incidence in the US (Clark et al. 2013). To reduce severe postharvest losses during storage and transport, roots are cured to allow for wound healing and toughening of the skin. Removal of infected roots and decontamination of packing equipment helps reduce secondary infections in the packing line and transport (Scruggs et al. 2017). Despite the implementation of these strategies in 2015, black rot remerged as a significant threat for the sweetpotato industry in North Carolina, Virginia, Georgia, Mississippi, Alabama, Louisiana, Arkansas, Missouri, and California. The 2015 outbreak prompted the approval of a crisis emergency label for using thiabendazole postharvest on sweetpotato in North Carolina (Quesada-Ocampo 2016). Scruggs et al. (2017) aimed to expand the control options by evaluating several chemicals that could potentially be labeled for sweetpotato use in the future. Based on their *in vitro* studies, difenoconazole was the most effective fungicide for inhibiting mycelial growth. This supports their observations on *in vivo* tests of Stadium (azoxystrobin + difenoconazole + fludioxonil) as the most effective postharvest fungicide in controlling black rot. Thiabendazole and fluazinam were also efficacious during *in vitro* and *in vivo* experiments.

As of 2018, North Carolina sweetpotato growers that intend to export sweetpotatoes to the European Union are encouraged to use thiabendazole only in field production. The European Union legislation has lowered the maximum residue level (MRL) for thiabendazole on sweetpotato to 0.01 ppm, which effectively eliminates the use of thiabendazole for postharvest disease management (Quesada-Ocampo 2018). The regulated use of thiabendazole limits the ability of sweetpotato growers to control black rot for international trade. In contrast, growers who intend to commercialize sweetpotatoes in the US, may count on an additional level of defense from a thiabendazole application prior to packing (Protection 2018).

Sweetpotato Resistance Against Ceratocystis fimbriata

Despite the availability of chemical control options, fungicide use is becoming limited due to changes in legislation and the evolution of pathogen resistance to chemicals (Michelmore et al. 2017). Deployment of host resistance represents a desirable tool to decrease crop losses due to plant pathogens. However, breeding sweetpotato for disease resistance in combination with high yield and aesthetic traits has been challenging due to the few breeding programs active in the US, the limited knowledge of the genetic basis of disease resistance in sweetpotato, and the hexaploid nature of the crop.

In 1953, a lack of knowledge regarding sources of resistance to black rot inspired a USDA plant pathologist to evaluate available germplasm (Cheo 1953). Cheo employed two methods of inoculation to evaluate black rot resistance in 100 sweetpotato varieties. First, he inoculated roots by introducing a *C. fimbriata* spore suspension in an 8-mm-deep injury and measured the lesion diameter. Then, he inoculated slips by dipping them into a spore suspension and transplanting them into the soil in a greenhouse setting. Slips were examined for below-ground black lesions. Cheo's experiments identified four varieties (B-5941, B-5944, B5999, B-6097) as highly resistant to black rot in both inoculation experiments. Four years later, Hildebrand (1957) assessed the levels of resistance in 300 sweetpotato selections by measuring the lesion diameter in roots. Hildebrand confirmed Cheo's results and identified B-5944, later named Sunnyside, as a positive control for future evaluations of host resistance. More recently, Scruggs et al. (2017) evaluated 14 common sweetpotato cultivars and advanced breeding lines for black rot resistance. Their results suggest that none of the cultivars were completely resistant to the disease and most were equally susceptible. Unfortunately, over the last 60 years, the resistance observed by Cheo and Hildebrand has not been bred into the currently available cultivars. This lack of complete resistance highlights the importance of examining sources of resistance in germplasm from the 1950s and wild relatives of sweetpotato to integrate the trait into commercially acceptable cultivars.

Plant Defenses Against Pathogens

Plants and pathogens exist in a continuum coevolutionary struggle for survival. In the context of global warming and disease triangle, climate conditions such as elevated temperatures and increased rainfall/humidity pose forces that change pathogen incidence, overwintering, emergence, and virulence, critically impacting food security (Velásquez et al. 2018). Despite all of that, plants are actually very good at fighting off pathogens, obeying the plant pathology dogma that most plants are resistant to most pathogens with disease representing the exception (Wu et al. 2018a). The main reason for how plants are able to ward off most pathogens is the fact that plants have a rapidly evolving immune system. Traditionally in plant pathology, the plant immune system has been described via the zig-zag model where four phases of plant immunity determine the outcome of the plant and microbe interactions (Jones and Dangl 2006). First, plants recognize pathogen-associated molecular patterns (PAMPs) such as flagellin in bacteria, glucans in oomycetes, chitin in fungi, ascarosides in nematodes, and coat protein in viruses. Such recognition is known as PAMP triggered immunity (PTI) and executed by pattern recognition receptors (PRRs) which are membrane bound receptors that contain an extracellular ligand-binding domain, a transmembrane domain, and an intracellular kinase domain. In the arms race for survival and colonization, pathogens evolved molecules termed "effectors" that can overcome PTI and result in a phase called effector triggered susceptibility (ETS). Subsequently, these pathogen secreted effectors can be recognized by intracellular immune receptors of which nucleotide-binding domain and leucine-rich repeat (NLR) proteins constitute one of the largest and most studied family of plant resistance proteins (Adachi et al. 2019a; Kourelis and van der Hoorn 2018). This process is termed effector triggered immunity (ETI) and leads to resistance in the absence of novel effectors. However, in the evolutionary arms race scenario, pathogens evolve to evade recognition or suppress ETI through novel effectors which in continued coevolution plants can restore ETI by evolving novel intracellular immune receptors (Cook et al. 2015).

Resistance proteins in a broad sense can be defined as genes that can provide full or partial resistance to one or more pathogens and these include PRRs and NLR proteins (Kourelis and van der Hoorn 2018). Here, we will focus on the latter, NLR intracellular immune receptors. Plant NLR architecture exhibits a multi-domain structure with conserved nucleotide-binding (NB) domain, C-terminal leucine rich repeat (LRR) domain, and diverse accessory N- and C- terminal domains as described by Jones et al. (2016). The majority of NLRs contain characteristic N terminal domains that can be themselves categorized into 3 groups, toll and interleukin-1 receptor (TIR)-type known as TNLs, coiled-coil (CC)-type known as CNLs, and RPW8-like CC type known as RNLs (Shao et al. 2016; Adachi et al. 2019a). Both CC and TIR domains appear to play key roles in oligomerization of NLR proteins (Bentham et al. 2017). Some CC domains appear to function as signaling domains capable of exerting hypersensitive response cell death through membrane perturbations (Wang et al. 2015; Adachi et al. 2019b; Wang et al. 2019). However, some CC domains are incapable of self-associating, highlighting the complexity that can be identified even at the domain and intramolecular level (Rairdan et al. 2008). The consensus is that the CC domain functions differently according to intermolecular connections in the NLR protein. TIR domains remain less characterized than CC domains and overall TIR domains are involved in activation and oligomerization (Monteiro and Nishimura 2018). The NB domain is believed to exchange adenosine diphosphate (ADP) for adenosine triphosphate (ATP), catalyzing the oligomerization of nod-like receptor (NLR) proteins and therefore enables signal amplification (Jones et al. 2016). Generally, LRR domains interact intramolecularly with diverse effector proteins and play a role in inhibition of NLR activity when pathogen effectors are missing (Balint-Kurti 2019; Monteiro and Nishimura 2018). While there is a pattern of structural conservancy and function, the mechanism for how NLR proteins suppress or activate immune response varies and there are always exceptions. That said in general the N-terminal domain seems to be involved in activation and downstream cell death pathways (Adachi et al. 2019a; Balint-Kurti 2019).

Next Generation Sequencing Accelerates Resistance Gene Discovery

Resistance Genes in Sweetpotato

Extensive examination of the literature revealed no functional characterization of predicted NLR genes in sweetpotato. The most recent attempt to annotate NLR genes from sweetpotato transcript sequences identified 94 genes with only 3 of them differentially expressed when inoculated with *F. oxysporum* f.sp. *batatas* (Lin et al. 2017). Their NLR gene annotation involved searches against common databases including the NR (NCBI non-redundant protein sequences), Pfam (Protein family), KOG/COG (Clusters of Orthologous Groups of proteins), and Blast2Go among others. Wang et al. (2010) attempted to isolate NLR genes from the sweetpotato cultivar Jewel by using degenerate primers based on conserved motifs within the NLR sequence. They identified 237 non-TIR and 38 TIR NLR sequences and stated the need of additional analysis to capture the full diversity of NLR genes in sweetpotato.

Genome Survey of NLR Encoding Genes in **I. batatas, I. trifida,** *and* **I. triloba**

Plants harbor hundreds of NLR encoding genes with extensive diversity and typically clustered as closely related paralogs (Jones et al. 2016). Their repetitive nature, extraordinary diversity and complexity, become an obstacle for generating accurate assemblies and annotations for NLR genes (Jones et al. 2016; Jupe et al. 2013; Yandell and Ence 2012). Annotation of NLR genes in plant genomes tends to be very laborious and time-consuming. Structural annotations of NLR genes are known to fail to capture full gene complements, leaving unannotated pieces or missing NLR domains in the final annotation files. Comparative genomic studies have exposed weaknesses of automated annotation pipelines, many of which tend to miss up to 50% of total NLR gene complement and/or split full sequences (Andolfo et al. 2014; Jupe et al. 2013; Meyers et al. 2003). Steuernagel et al. (2015) released a robust and powerful NLR annotation tool known as NLR-parser. NLR-parser allows for rapid annotation of plant NLR genes by using the highly specific amino acid motif structure found in NLR gene products. The output of this program is an easy-to-use tabular file with information about the completeness and type of NLR gene present in the genome.

In a proof of concept study, we compared the original genome annotation of 18 plant species against the NLR-parser annotation output. This comparison included the genomes of *Ananas comosus* v3.0 (Ming et al. 2015), *Arabidopsis thaliana* TAIR10 (Lamesch et al. 2012), *Chenopodium quinoa* v1.0 (Jarvis et al. 2017), *Cucumis sativus* v1.0 (Phytozome 2012), *Eucalyptus grandis* v2.0 (Myburg et al.

2014), *I. trifida* v3.0, *I. littoralis* v3.0 (Wu et al. 2018b), *Lactuca sativa* v8.0 (Reyes-Chin-Wo et al. 2017), *Manihot esculenta* v6.1 (Bredeson et al. 2016), *Musa acuminata* v1.0 (Droc et al. 2013), *Oryza sativa* v7.0 (Ouyang et al. 2006), *Sorghum bicolor* v3.1.1 (McCormick et al. 2018), *S. chacoense* v1.0 (Leisner et al. 2018), *S. tuberosum* v4.03 (Xu et al. 2011), *Theobroma cacao* v2 (Motamayor et al. 2013), *Vitis vinifera* v2.0 (Jaillon et al. 2007), and *Zea mays* v4 (Schnable et al. 2009). In general, we observed a pattern of misannotation of NLR genes across all 18 genomes compared here. Interestingly, wild relatives of sweetpotato (*I. trifida*, *I. triloba*) and white potato (*S. chacoense*) exhibited higher NLR counts in their respective genome annotation (Fig. 3.3). The same research group was involved in the assembly and annotation of these 3 wild relatives. Their custom pipeline used for functionally annotating gene products seems to capture a greater number of NLRs; however, a few differences were observed for false positives and negatives. Our comparative genome analysis emphasizes the need to improve sequencing, assembly, and annotation of the NLR gene family as critical for studies aiming to breed for resistance to plant pathogens. Interestingly, Xu et al. (2011) employed a more specific approach to annotate NLRs in the potato genome version 4 including an HMMER (http://hmmer.janelia.org/software) motif search to identify the NB-ARC domain, multiple expectation maximization for motif elicitation (MEME) (Bailey and Elkan 1995) to reveal the LLR and TIR domains, and MARCOIL (Delorenzi and Speed 2002) to detect CC domains. This annotation pipeline resulted in the annotation of 148 NLRs. In contrast, using NLR parser, we were able to annotate 726 NLRs as complete or partial. Despite the identification of NLRs in our study, a large number of NLRs were categorized by NLR-parser as partial, meaning at least one NLR domain was identified, stressing the difficulty to sequence and assemble NLRs. Genome projects or assemblers often fail to capture the entire length of the NLR complement. Fortunately, target sequence capture combined with long read DNA sequencing technology, a technique known as RenSeq (Resistance gene enrichment sequencing), now allows for the generation of complete NLR gene repertoires in agronomically important crops (Giolai et al. 2017; Jupe et al. 2013; Witek et al. 2016). Taking advantage of biotinylated RNA probes designed to capture 450 predicted NLR genes, Jupe et al. (2013) sequenced 750 NLR genes in *S. tuberosum*. RenSeq reduces the complexity of the genome through a target enrichment approach producing sequences for the NLR gene family. Currently, RenSeq is also being exploited to identify the resistome of crop wild relatives (Steuernagel et al. 2016). Stam et al. (2016) showed that RenSeq can be used to identify 220 NLR genes in a single population of *S. pennellii*, a wild relative of cultivated tomatoes.

We built a phylogeny of complete NLRs from Solanaceae and Convolvulaceae sequenced genomes to test the relationship of NLRs in these two plant families (Fig. 3.4). This phylogeny included 15 well-characterized NLRs from a curated NLR database (Sanseverino et al. 2012). The coil-coiled NLRs, also known as CNLs, tend to dominate the phylogeny. Our NLR phylogeny confirms the existence of a well-supported superclade that includes well-known NLRs, such as Rpi-blb2, Mi-1.2, Hero, Sw5-a and R1, all of which confer resistance to diverse plant pathogens, and is a highly ancient clade that emerged 100 million years ago (Wu et al.

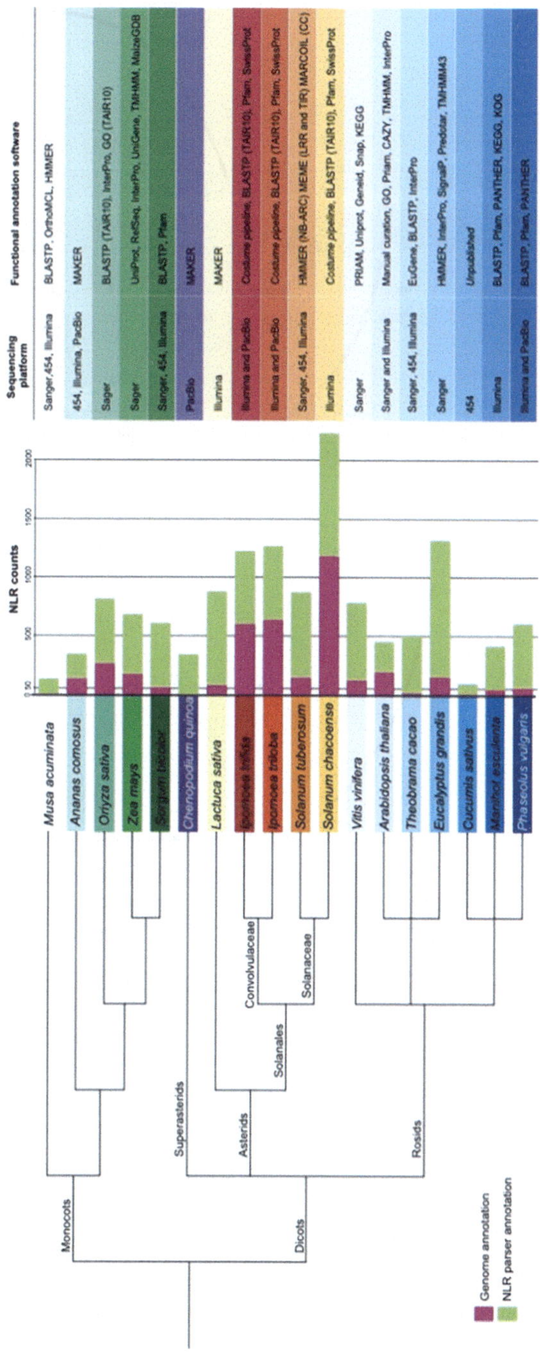

Fig. 3.3 Summary of phylogeny and NLR counts as annotated by genome projects annotation pipelines and NLR-parser in different plant species. A phylogenetic tree of plant species was built using phyloT based on National Center for Biotechnology Information taxon identification numbers. Sequencing platforms and respective functional gene annotation softwares used for each plant species genome project are listed

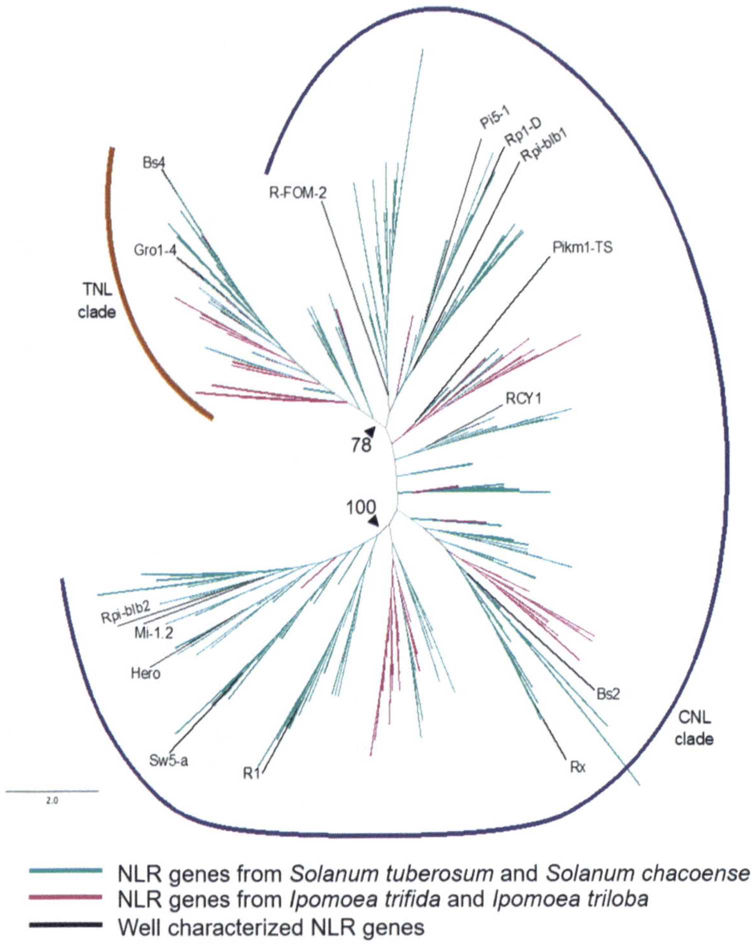

Fig. 3.4 NLR unrooted phylogenetic tree for *Solanum tuberosum*, *Solanum chacoense*, *Ipomoea trifida*, *Ipomoea triloba* and 15 well-characterized NLRs from a curated database (Sanseverino et al. 2012). Phylogenetic tree based on the NB-ARC1-ARC2 domain were generated using IQTREE v.1.6.10 (Hoang et al. 2017). Numbers at branches indicate bootstrap values for the branching of the major clades. Peripheral label indicates clade designation. Scale bar indicates 2.0 average amino acid substitutions per site

2017). Interestingly only one NLR from the Convolvulaceae belongs to that super-clade, implying differences in how this gene family perhaps evolved in the Solanales order. This small phylogeny could potentially be improved with annotated NLRs from cultivated sweetpotato and used to study evolution of NLRs from diploid wild relatives to hexaploid or tetraploid cultivated sweetpotatoes and white potatoes respectively.

Future Perspectives

Achieving disease resistance of sweetpotato requires the identification and recruitment of large numbers of diverse *R* genes (Michelmore et al. 2017). Examining the sweetpotato genome to reveal resistant candidate genes can greatly benefit the sweetpotato industry worldwide, and in particular in the US, but also farmers in developing economies where crop losses due to plant pathogens are devastating subsistence farming. As agriculture moves towards lower input and sustainable practices under a changing climate, it becomes critical to strategize potential scenarios where pathogens spread into new areas or re-emerge more aggressively. A resistance gene repertoire represents an adaptable management tool to develop climate resilient crops. To translate the resistome of sweetpotato into valuable breeding tools, it is imperative to combine resistome-based marker assisted selection with high throughput phenotyping of segregating populations as well as expanding the knowledge of plant pathogen genomics to be ahead of the coevolutionary dynamics of host-pathogen interactions.

References

Adachi H, Derevnina L, Kamoun S (2019a) NLR singletons, pairs, and networks: evolution, assembly, and regulation of the intracellular immunoreceptor circuitry of plants. Curr Opin Plant Biol 50:121–131. https://doi.org/10.1016/j.pbi.2019.04.007

Adachi H, Kamoun S, Maqbool A (2019b) A resistosome-activated 'death switch. Nat Plants 5:457–458. https://doi.org/10.1038/s41477-019-0425-9

Andolfo G, Jupe F, Witek K, Etherington GJ, Ercolano MR, Jones JDG (2014) Defining the full tomato NB-LRR resistance gene repertoire using genomic and cDNA RenSeq. BMC Plant Biol 14:120. https://doi.org/10.1186/1471-2229-14-120

Austin DF (1987) The taxonomy, evolution and genetic diversity of sweet potatoes and related wild species. In: Gregory P (ed) Exploration, maintenance, utilization of sweet potato genetic resources. Proceedings of the first planning conference. International Potato Center (CIP), Lima, pp 25–59

Bailey TL, Elkan C (1995) The value of prior knowledge in discovering motifs with MEME. Proc Int Conf Intelligent Syst Mol Biol 3:21–29

Balint-Kurti P (2019) The plant hypersensitive response: concepts, control and consequences. Mol Plant Pathol 20:1163–1178. https://doi.org/10.1111/mpp.12821

Bentham A, Burdett H, Anderson PA, Williams SJ, Kobe B (2017) Animal NLRs provide structural insights into plant NLR function. Ann Bot 119:827–702. https://doi.org/10.1093/aob/mcw171

Bertioli DJ, Cannon SB, Froenicke L, Huang G, Farmer AD, Cannon EK, Liu X, Gao D, Clevenger J, Dash S, Ren L (2016) The genome sequences of *Arachis duranensis* and *Arachis ipaensis*, the diploid ancestors of cultivated peanut. Nat Gen 48:438–446. https://doi.org/10.1038/ng.3517

Boudreaux JE (2005) Commercial vegetable production recommendations. (Louisiana Cooperative Extension Service)(USA)

Bradshaw J (2010) Root and tuber crops. In: Handbook of plant breeding. Springer, New York. https://doi.org/10.1007/978-0-387-92765-7

Bredeson J, Lyons J, Prochnik S, Wu G, Ha C, Edsinger-Gonzales E, Grimwood J, Schmutz J, Rabbi I, Egesi C, Nauluvula P, Lebot V, Ndunguru J, Mkamilo G, Bart R, Setter T, Gleadow R, Kulakow P, Ferguson M, Rounsley S, Rokhsar D (2016) Sequencing wild and cultivated cas-

sava and related species reveals extensive interspecific hybridization and genetic diversity. Nat Biotechnol 34:562–570. https://doi.org/10.1038/nbt.3535
Cheo PC (1953) Varietal differences in susceptibility of sweetpotato to black rot fungus. Phytopathology 43:78–81
CIP (2010) Fact and figures about sweetpotato. https://cipotato.org/sweetpotato/sweetpotato-facts-and-figures/. Accessed 23 Apr 2020
Clark CA (1992) Post harvest diseases of sweetpotatoes and their control. Postharvest News Inf 3:4
Clark CA, Davis JA, Abad JA, Cuellar WJ, Fuentes S, Kreuze J, Gibson RW, Mukasa SB, Tugume AK, Tairo FD, Valkonen J (2012) Sweetpotato viruses: 15 years of progress on understanding and managing complex diseases. Plant Dis 96:168–185
Clark CA, Ferrin DM, Smith TP, Holmes GJ (2013) Compendium of sweetpotato diseases, pests, and disorders, 2nd edn. American Phytopathological Society, St. Paul
Clarke AC, Burtenshaw MK, McLenachan PA, Erickson DL, Penny D (2006) Reconstructing the origins and dispersal of the Polynesian bottle gourd (*Lagenaria siceraria*). Mol Biol Evol 23:893–900. https://doi.org/10.1093/molbev/msj092
Cook DE, Mesarich CH, Thomma BPHJ (2015) Understanding plant immunity as a surveillance system to detect invasion. Annu Rev Phytopathol 53:541–563. https://doi.org/10.1146/annurev-phyto-080614-120114
Delorenzi M, Speed T (2002) An HMM model for coiled-coil domains and a comparison with PSSM-based predictions. Bioinformatics 18:617–625. https://doi.org/10.1093/bioinformatics/18.4.617
Droc G, Larivière D, Guignon V, Yahiaoui N, This D, Garsmeur O, Dereeper A, Hamelin C, Argout X, Dufayard J, Lengelle J, Baurens F, Cenci A, Pitollat B, D'Hont A, Ruiz M, Rouard M, Bocs S (2013) The banana genome hub. Database 2013:1–14
Edmunds B, Boyette M, Clark CA, Ferrin DM, Smith TP, Holmes GJ (2008) Postharvest handling of sweetpotatoes. NC State Extension Publications
Engelbrecht CJB (2004) Host specialization, intersterility, and taxonomy of populations of *Ceratocystis fimbriata* from sweet potato, sycamore, and cacao. Iowa State University
FAOSTAT (2016) Exports quantity for Sweetpotatoes. http://www.fao.org/faostat/en/#data/TP
FAOSTAT (2017a) Low-Income Food-Deficit Countries (LIFDCs) – list for 2018. http://www.fao.org/countryprofiles/lifdc/en/. Accessed 23 Apr 2020
FAOSTAT (2017b) Sweetpotato production. http://www.fao.org/faostat/en/#search/Sweet%20potatoes. Accessed 23 Apr 2020
Giolai M, Paajanen P, Verweij W, Witek K, Jones JDG, Clark MD (2017) Comparative analysis of targeted long read sequencing approaches for characterization of a plant's immune receptor repertoire. BMC Genomics 18:564. https://doi.org/10.1186/s12864-017-3936-7
Halsted BD (1890) Some fungous diseases of the sweet potato, Bulleting no. 76. Agriculture college Experiment Station, New Brunswick
Hildebrand EM (1957) Testing sweetpotatoes for black rot resistance. Plant Dis Rep 41:661–670
Hirakawa H, Okada Y, Tabuchi H, Shirasawa K, Watanabe A, Tsuruoka H, Minami C, Nakayama S, Sasamoto S, Kohara M, Kishida Y, Fujishiro T, Kato M, Nanri K, Komaki A, Yoshinaga M, Takahata Y, Tanaka M, Tabata S, Isobe SN (2015) Survey of genome sequences in a wild sweet potato, *Ipomoea trifida* (H. B. K.) G. Don. DNA Res 22:171–179. https://doi.org/10.1093/dnares/dsv002
Hoang DT, Chernomor O, Av H, Minh BQ, Vinh LS (2017) UFBoot2: improving the ultrafast bootstrap approximation. Mol Biol Evol 35(2):518–522. https://doi.org/10.1093/molbev/msx281
Holmes G, Kemble J (2010) Southeastern US vegetable crop handbook. North Carolina State Cooperative Extension Service, Raleigh
Isobe S, Shirasawa K, Hirakawa H (2017) Challenges to genome sequence dissection in sweetpotato. Breeding Sci 67:35–40. https://doi.org/10.1270/jsbbs.16186
Jaillon O, Aury J, Noel B, Policriti A, Clepet C, Casagrande A, Choisne N, Aubourg S, Vitulo N, Jubin C, Vezzi A, Legeai F, Hugueney P, Dasilva C, Horner D, Mica E, Jublot D, Poulain J, Bruyère C, Billault A, Segurens B, Gouyvenoux M, Ugarte E, Cattonaro F, Anthouard V, Vico

V, Del FC, Alaux M, Di GG, Dumas V, Felice N, Paillard S, Juman I, Moroldo M, Scalabrin S, Canaguier A, Le CI, Malacrida G, Durand E, Pesole G, Laucou V, Chatelet P, Merdinoglu D, Delledonne M, Pezzotti M, Lecharny A, Scarpelli C, Artiguenave F, Pè M, Valle G, Morgante M, Caboche M, Adam-Blondon A, Weissenbach J, Quétier F, Wincker P (2007) The grapevine genome sequence suggests ancestral hexaploidization in major angiosperm phyla. Nature 449:463–467. https://doi.org/10.1038/nature06148

Jarvis D, Ho Y, Lightfoot D, Schmöckel S, Li B, Borm T, Ohyanagi H, Mineta K, Michell C, Saber N, Kharbatia N, Rupper R, Sharp A, Dally N, Boughton B, Woo Y, Gao G, Schijlen E, Guo X, Momin A, Negrão S, Al-Babili S, Gehring C, Roessner U, Jung C, Murphy K, Arold S, Gojobori T, Linden C, van L, EN JE, Maughan P, Tester M (2017) The genome of *Chenopodium quinoa*. Nature 542:307–311. https://doi.org/10.1038/nature21370

Jones JDG, Dangl JL (2006) The plant immune system. Nature 444:323–329. https://doi.org/10.1038/nature05286

Jones JDG, Vance RE, Dangl JL (2016) Intracellular innate immune surveillance devices in plants and animals. Science 354:aaf6395. https://doi.org/10.1126/science.aaf6395

Jupe F, Witek K, Verweij W, Śliwka J, Pritchard L, Etherington GJ, MacLean D, Cock PJ, Leggett RM, Bryan GJ, Cardle L, Hein I, Jones JDG (2013) Resistance gene enrichment sequencing (RenSeq) enables reannotation of the NB-LRR gene family from sequenced plant genomes and rapid mapping of resistance loci in segregating populations. Plant J 76:530–544. https://doi.org/10.1111/tpj.12307

Karna P, Gundala SR, Gupta MV, Shamsi SA, Pace RD, Yates C, Narayan S, Aneja R (2011) Polyphenol-rich sweet potato greens extract inhibits proliferation and induces apoptosis in prostate cancer cells in vitro and in vivo. Carcinogenesis 32:1872–1880. https://doi.org/10.1093/carcin/bgr215

Kokkinos CD, Clark CA (2006) Real-time PCR assays for detection and quantification of sweetpotato viruses. Plant Dis 90:783–788. https://doi.org/10.1094/PD-90-0783

Kourelis J, van der Hoorn RAL (2018) Defended to the nines: 25 years of resistance gene cloning identifies nine mechanisms for R protein function. Plant Cell 30:285–299. https://doi.org/10.1105/tpc.17.00579

Lamesch P, Berardini TZ, Li D, Swarbreck D, Wilks C, Sasidharan R, Muller R, Dreher K, Alexander DL, Garcia-Hernandez M, Karthikeyan AS, Lee CH, Nelson WD, Ploetz L, Singh S, Wensel A, Huala E (2012) The Arabidopsis Information Resource (TAIR): improved gene annotation and new tools. Nucleic Acids Res 40:1202–1210. https://doi.org/10.1093/nar/gkr1090

Lebot V (2010) Sweet potato. In: Root and tuber crops. Springer, Berlin, pp 97–125

Leisner C, Hamilton J, Crisovan E, Manrique-Carpintero N, Marand A, Newton L, Pham G, Jiang J, Douches D, Jansky S, Buell C (2018) Genome sequence of M6, a diploid inbred clone of the high-glycoalkaloid-producing tuber-bearing potato species *Solanum chacoense*, reveals residual heterozygosity. Plant J 94:562–570. https://doi.org/10.1111/tpj.13857

Lin Y, Zou W, Lin S, Onofua D, Yang Z, Chen H, Wang S, Chen X (2017) Transcriptome profiling and digital gene expression analysis of sweet potato for the identification of putative genes involved in the defense response against *Fusarium oxysporum* f. sp. *batatas*. PLoS One 12:e0187838. https://doi.org/10.1371/journal.pone.0187838

Ling H-Q, Zhao S, Liu D, Wang J, Sun H, Zhang C, Fan H, Li D, Dong L, Tao Y (2013) Draft genome of the wheat A-genome progenitor *Triticum urartu*. Nature 496:87–90

Loebenstein G (2009) Origin, distribution and economic importance. In: Loebenstein G, Thottappilly G (eds) The Sweetpotato. Springer, Berlin, pp 9–12. https://doi.org/10.1007/978-1-4020-9475-02

Loebenstein G, Thottappilly G (2009) The sweetpotato. Springer Science & Business Media, Dordrecht

McCormick R, Truong S, Sreedasyam A, Jenkins J, Shu S, Sims D, Kennedy M, Amirebrahimi M, Weers B, McKinley B, Mattison A, Morishige D, Grimwood J, Schmutz J, Mullet J (2018) The *Sorghum bicolor* reference genome: improved assembly, gene annotations, a transcriptome atlas, and signatures of genome organization. Plant J 93:338–354. https://doi.org/10.1111/tpj.13781

Meyers BC, Kozik A, Griego A, Kuang H, Michelmore RW (2003) Genome-wide analysis of NBS-LRR-encoding genes in Arabidopsis. Plant Cell 15:809–834. https://doi.org/10.1105/tpc.009308

Michael TP, Jackson S (2013) The first 50 plant genomes. Plant Genome 6:1–7. https://doi.org/10.3835/plantgenome2013.03.0001in

Michael TP, VanBuren R (2015) Progress, challenges and the future of crop genomes. Curr Opin Plant Biol 24:71–81. https://doi.org/10.1016/j.pbi.2015.02.002

Michelmore RW, Coaker G, Bart R, Beattie GA, Bent A, Bruce T, Cameron D, Dangl J, Dinesh-Kumar S, Edwards R, den Akker S, Gassmann W, Greenberg J, Harrison R, He P, Harvey J, Huffaker A, Hulbert S, Innes R, Jones JD, Kaloshian I, Kamoun S, Katagiri F, Leach JE, Ma W, McDowell JM, Medford J, Meyers B, Nelson R, Oliver R, Qi Y, Saunders D, Shaw M, Subudhi P, Torrance L, Tyler BM, Walsh J (2017) Foundational and translational research opportunities to improve plant health. MPMI 30:1–37. https://doi.org/10.1094/MPMI-01-17-0010-CR

Ming R, VanBuren R, Wai CM, Tang H, Schatz MC, Bowers JE, Lyons E, Wang M-LL, Chen J, Biggers E, Zhang J, Huang L, Zhang L, Miao W, Zhang J, Ye Z, Miao C, Lin Z, Wang H, Zhou H, Yim WC, Priest HD, Zheng C, Woodhouse M, Edger PP, Guyot R, Guo H-BB, Guo H, Zheng G, Singh R, Sharma A, Min X, Zheng Y, Lee H, Gurtowski J, Sedlazeck FJ, Harkess A, McKain MR, Liao Z, Fang J, Liu J, Zhang X, Zhang Q, Hu W, Qin Y, Wang K, Chen L-YY, Shirley N, Lin Y-RR, Liu L-YY, Hernandez AG, Wright CL, Bulone V, Tuskan GA, Heath K, Zee F, Moore PH, Sunkar R, Leebens-Mack JH, Mockler T, Bennetzen JL, Freeling M, Sankoff D, Paterson AH, Zhu X, Yang X, Smith JA, Cushman JC, Paull RE, Yu Q (2015) The pineapple genome and the evolution of CAM photosynthesis. Nat Gen 47:1435–1442. https://doi.org/10.1038/ng.3435

Monteiro F, Nishimura MT (2018) Structural, functional, and genomic diversity of plant NLR proteins: an evolved resource for rational engineering of plant immunity. Annu Rev Phytopathol 56:243–267. https://doi.org/10.1146/annurev-phyto-080417-045817

Motamayor J, Mockaitis K, Schmutz J, Haiminen N, Livingstone D, Cornejo O, Findley S, Zheng P, Utro F, Royaert S, Saski C, Jenkins J, Podicheti R, Zhao M, Scheffler B, Stack J, Feltus F, Mustiga G, Amores F, Phillips W, Marelli J, May G, Shapiro H, Ma J, Bustamante C, Schnell R, Main D, Gilbert D, Parida L, Kuhn D (2013) The genome sequence of the most widely cultivated cacao type and its use to identify candidate genes regulating pod color. Genome Biol 14:r53. https://doi.org/10.1186/gb-2013-14-6-r53

Myburg A, Grattapaglia D, Tuskan G, Hellsten U, Hayes R, Grimwood J, Jenkins J, Lindquist E, Tice H, Bauer D, Goodstein D, Dubchak I, Poliakov A, Mizrachi E, Kullan A, Hussey S, Pinard D, Karen Van Der Merwe, Singh P, Ida Van Jaarsveld, Silva-Junior O, Togawa R, Pappas M, Faria D, Sansaloni C, Petroli C, Yang X, Ranjan P, Tschaplinski T, Ye C, Li T, Sterck L, Vanneste K, Murat F, Soler M, Clemente H, Saidi N, Cassan-Wang H, Dunand C, Hefer C, Bornberg-Bauer E, Kersting A, Vining K, Amarasinghe V, Ranik M, Naithani S, Elser J, Boyd A, Liston A, Spatafora J, Dharmwardhana P, Raja R, Sullivan C, Romanel E, Alves-Ferreira M, Külheim C, Foley W, Carocha V, Paiva J, Kudrna D, Brommonschenkel S, Pasquali G, Byrne M, Rigault P, Tibbits J, Spokevicius A, Jones R, Steane D, Vaillancourt R, Potts B, Joubert F, Barry K, Pappas G, Strauss S, Jaiswal P, Grima-Pettenati J, Salse J, Yves Van De Peer, Daniel S. Rokhsar, Schmutz J (2014) The genome of *Eucalyptus grandis*. Nature 510:356-362. https://doi.org/10.1038/nature13308

O'Brien PJ (1972) The sweet potato: its origin and dispersal. Am Anthropol 74:342–365. https://doi.org/10.1525/aa.1972.74.3.02a00070

Ouyang S, Zhu W, Hamilton J, Lin H, Campbell M, Childs K, Thibaud-Nissen F, Malek R, Lee Y, Zheng L, Orvis J, Haas B, Wortman J, Buell C (2006) The TIGR rice genome annotation resource: improvements and new features. Nucleic Acids Res 35:883–887. https://doi.org/10.1093/nar/gkl976

Ozias-Akins P, Jarret RL (1994) Nuclear DNA content and ploidy levels in the genus *Ipomoea*. J Am Soc Hortic Sci 119:110–115. https://doi.org/10.21273/JASHS.119.1.110
Phytozome (2012) *Cucumis sativus* v1.0 (Cucumber). DOE. https://phytozome.jgi.doe.gov/pz/portal.html#!info?alias=Org_Csativus
Protection SC (2018) Section 18 emergency exemption for postharvest treatment of sweet potato for control of blackrot. 18NC01
Quesada-Ocampo LM (2016) Emergency section 18 label approved for mertect to control black rot postharvest on sweetpotato in North Carolina. Extension Plant Pathology
Quesada-Ocampo LM (2018) Section 18 label renewed in North Carolina for use of mertect post-harvest to control black rot of sweetpotato for domestic markets. Extension Plant Pathology
Rairdan GJ, Collier SM, Sacco MA, Baldwin TT, Boettrich T, Moffett P (2008) The coiled-coil and nucleotide binding domains of the potato Rx disease resistance protein function in pathogen recognition and signaling. Plant Cell 20:739–751. https://doi.org/10.1105/tpc.107.056036
Reyes-Chin-Wo S, Wang Z, Yang X, Kozik A, Arikit S, Song C, Xia L, Froenicke L, Lavelle D, Truco M, Xia R, Zhu S, Xu C, Xu H, Xu X, Cox K, Korf I (2017) Genome assembly with *in vitro* proximity ligation data and whole-genome triplication in lettuce. Nat Commun 8:14953. https://doi.org/10.1038/ncomms14953
Roullier C, Rossel G, Tay D, McKey D (2011) Combining chloroplast and nuclear microsatellites to investigate origin and dispersal of New World sweet potato landraces. Mol Ecol 20:3963–3977. https://doi.org/10.1111/j.1365-294X.2011.05229.x
Roullier C, Benoit L, McKey DB, Lebot V (2013) Historical collections reveal patterns of diffusion of sweet potato in Oceania obscured by modern plant movements and recombination. Proc Natl Acad Sci USA 110:2205–2210. https://doi.org/10.1073/pnas.1211049110
Saigusa N, Terahara N, Riichiro OHBA (2005) Evaluation of DPPH-radical-scavenging activity and antimutagenicity and analysis of anthocyanins in an alcoholic fermented beverage produced from cooked or raw purple-fleshed sweet potato (*Ipomoea Batatas* cv. Ayamurasaki) roots. Food Sci Technol Res 11:390–394. https://doi.org/10.3136/fstr.11.390
Sanseverino W, Hermoso A, D'alessandro R, Vlasova A, Andolfo G, Frusciante L, Lowy E, Roma G, Ercolano M (2012) PRGdb 2.0: towards a community-based database model for the analysis of R-genes in plants. Nucleic Acids Res 41:1167–1171. https://doi.org/10.1093/nar/gks1183
Schnable PS, Ware D, Fulton RS, Stein JC, Wei F, Pasternak S, Liang C, Zhang J, Fulton L, Graves TA, Minx P, Reily AD, Courtney L, Kruchowski SS, Tomlinson C, Strong C, Delehaunty K, Fronick C, Courtney B, Rock SM, Belter E, Du F, Kim K, Abbott RM, Cotton M, Levy A, Marchetto P, Ochoa K, Jackson SM, Gillam B, Chen W, Yan L, Higginbotham J, Cardenas M, Waligorski J, Applebaum E, Phelps L, Falcone J, Kanchi K, Thane T, Scimone A, Thane N, Henke J, Wang T, Ruppert J, Shah N, Rotter K, Hodges J, Ingenthron E, Cordes M, Kohlberg S, Sgro J, Delgado B, Mead K, Chinwalla A, Leonard S, Crouse K, Collura K, Kudrna D, Currie J, He R, Angelova A, Rajasekar S, Mueller T, Lomeli R, Scara G, Ko A, Delaney K, Wissotski M, Lopez G, Campos D, Braidotti M, Ashley E, Golser W, Kim H, Lee S, Lin J, Dujmic Z, Kim W, Talag J, Zuccolo A, Fan C, Sebastian A, Kramer M, Spiegel L, Nascimento L, Zutavern T, Miller B, Ambroise C, Muller S, Spooner W, Narechania A, Ren L, Wei S, Kumari S, Faga B, Levy MJ, McMahan L, Van Buren P, Vaughn MW (2009) The B73 maize genome: complexity, diversity, and dynamics. Science 326:1112–1115. https://doi.org/10.1126/science.1178534
Scruggs AC, Quesada-Ocampo LM (2016) Etiology and epidemiological conditions promoting Fusarium root rot in sweetpotato. Phytopathology 106:909–919. https://doi.org/10.1094/PHYTO-01-16-0009-R
Scruggs AC, Basaiah T, Adams ML, Quesada-Ocampo LM (2017) Genetic diversity, fungicide sensitivity, and host resistance to *Ceratocystis fimbriata* infecting sweetpotato in North Carolina. Plant Dis 101:994–1001. https://doi.org/10.1094/PDIS-11-16-1583-RE
Shao Z-Q, Xue J-Y, Wu P, Zhang Y-M, Wu Y, Hang Y-Y, Wang B, Chen J-Q (2016) Large-scale analyses of angiosperm nucleotide-binding site-leucine-rich repeat genes reveal three anciently diverged classes with distinct evolutionary patterns. Plant Physiol 170:2095–2109. https://doi.org/10.1104/pp.15.01487

Sheikha EAF, Ray RC (2017) Potential impacts of bioprocessing of sweet potato. Crit Rev Food Sci 57:455–471. https://doi.org/10.1080/10408398.2014.960909
Smith T, Stoddard S, Shankle M, Schultheis J (2009) Sweetpotato production in the United States. In: The sweetpotato. Springer, pp 287–323
Stam R, Scheikl D, Tellier A (2016) Pooled enrichment sequencing identifies diversity and evolutionary pressures at NLR resistance genes within a wild tomato population. Genome Biol Evol 8:1501–1515. https://doi.org/10.1093/gbe/evw094
Steuernagel B, Jupe F, Witek K, Jones JD, Wulff BB (2015) NLR-parser: rapid annotation of plant NLR complements. Bioinformatics 31:1665–1667. https://doi.org/10.1093/bioinformatics/btv005
Steuernagel B, Periyannan SK, Hernández-Pinzón I, Witek K, Rouse MN, Yu G, Hatta A, Ayliffe M, Bariana H, Jones JDG, Lagudah ES, Wulff BBH (2016) Rapid cloning of disease-resistance genes in plants using mutagenesis and sequence capture. Nat Biotechnol 34:652–655. https://doi.org/10.1038/nbt.3543
USDA (2016) Sweetpotato insurance standards handbook. USDA
USDA (2017) National nutrient database for standard reference release 28: sweet potato. USDA
USDA-ERS (2018) Sweet potatoes: US exports value and volume. USDA-ERS
USDA-NASS (2018) Sweet potato production measured in dollars. USDA-NASS
Velásquez AC, Castroverde CDM, He SY (2018) Plant–pathogen warfare under changing climate conditions. Curr Biol 28:619–634. https://doi.org/10.1016/j.cub.2018.03.054
Villordon AQ, Ginzberg I, Firon N (2014) Root architecture and root and tuber crop productivity. Trends Plant Sci 19(7):419–425. https://doi.org/10.1016/j.tplants.2014.02.002
Wang LS, D. Stoner G (2008) Anthocyanins and their role in cancer prevention. Cancer Lett. 269:281-290
Wang Y, Rosen B, Scoffield J, Egnin M, Mortley D, Steiner S, Cook DR, He G (2010) Isolation and analysis of resistance gene homologues in sweetpotato. Plant Breed 129:519–525. https://doi.org/10.1111/j.1439-0523.2009.01711.x
Wang G-F, Ji J, Ei-Kasmi F, Dangl JL, Johal G, Balint-Kurti PJ (2015) Molecular and functional analyses of a maize autoactive NB-LRR protein identify precise structural requirements for activity. PLoS Pathog 11:e1004674. https://doi.org/10.1371/journal.ppat.1004674
Wang J, Hu M, Wang J, Qi J, Han Z, Wang G, Qi Y, Wang H-WW, Zhou J-MM, Chai J (2019) Reconstitution and structure of a plant NLR resistosome conferring immunity. Science 364:eaav5870. https://doi.org/10.1126/science.aav5870
Witek K, Jupe F, Witek AI, Baker D, Clark MD (2016) Accelerated cloning of a potato late blight-resistance gene using RenSeq and SMRT sequencing. Nat Biotechnol 34:656–660. https://doi.org/10.1038/nbt.3540
Wu C-HH, Abd-El-Haliem A, Bozkurt TO, Belhaj K, Terauchi R, Vossen JH, Kamoun S (2017) NLR network mediates immunity to diverse plant pathogens. Proc Natl Acad Sci USA 114:8113–8118. https://doi.org/10.1073/pnas.1702041114
Wu C-H, Derevnina L, Kamoun S (2018a) Receptor networks underpin plant immunity. Science 360:1300–1301. https://doi.org/10.1126/science.aat2623
Wu S, Lau KH, Cao Q, Hamilton JP, Sun H, Zhou C, Eserman L, Gemenet DC, Olukolu BA, Wang H (2018b) Genome sequences of two diploid wild relatives of cultivated sweetpotato reveal targets for genetic improvement. Nat Commun 9:4580. https://doi.org/10.1038/s41467-018-06983-8
Xu X, Pan S, Cheng S, Zhang B, Mu D, Ni P, Zhang G, Yang S, Li R, Wang J, Orjeda G, Guzman F, Torres M, Lozano R, Ponce O, Martinez D, De la Cruz G, Chakrabarti SK, Patil VU, Skryabin KG, Kuznetsov BB, Ravin NV, Kolganova TV, Beletsky AV, Mardanov AV, Di Genova A, Bolser DM, Martin DMA, Li G, Yang Y, Kuang H, Hu Q, Xiong X, Bishop GJ, Sagredo B, Mejía N, Zagorski W, Gromadka R, Gawor J, Szczesny P, Huang S, Zhang Z, Liang C, He J, Li Y, He Y, Xu J, Zhang Y, Xie B, Du Y, Qu D, Bonierbale M, Ghislain M, del Rosario Herrera M, Giuliano G, Pietrella M, Perrotta G, Facella P, O'Brien K, Feingold SE, Barreiro LE, Massa GA, Diambra L, Whitty BR, Vaillancourt B, Lin H, Massa AN, Geoffroy M, Lundback S,

DellaPenna D, Robin Buell C, Sharma SK, Marshall DF, Waugh R, Bryan GJ, Destefanis M, Nagy I, Milbourne D, Thomson SJ, Fiers M, Jacobs JME, Nielsen KL, Sønderkær M, Iovene M, Torres GA, Jiang J, Veilleux RE, Bachem CWB, de Boer J, Borm T, Kloosterman B, van Eck H, Datema E, te Lintel Hekkert B, Goverse A, van Ham RCHJ, Visser RGF, The Potato Genome Sequencing C, The Potato Genome C, Shenzhen BGI, Cayetano Heredia U, Central Potato Research I, Centre Bioengineering RAS, Cgr-Cmm UdC, College of Life Sciences UoD, High Technology Research Center SAoAS, Huazhong Agriculture U, Hunan Agricultural U, Imperial College L, Instituto de Investigaciones A, Institute of B, Biophysics, Institute of V, Flowers CAoAS, International Potato C, Italian National Agency for New Technologies E, Sustainable D, Institute JCV, Laboratorio de Agrobiotecnología INdTA, Laboratorio de Biología de Sistemas UNdL, Michigan State U, Scottish Crop Research I, Teagasc Crops Research C, The New Zealand Institute for P, Food Research L, University of A, University of W, Virginia Polytechnic I, State U, Wageningen U, Research C (2011) Genome sequence and analysis of the tuber crop potato. Nature 475:189–195. https://doi.org/10.1038/nature10158

Yan Q, Vang LVV, Khanh CN, Naka H, Ando T (2014) Reexamination of the female sex pheromone of the sweet potato vine borer moth: identification and field evaluation of a tricosatriene. J Chem Ecol 40:590–598. https://doi.org/10.1007/s10886-014-0446-9

Yandell M, Ence D (2012) A beginner's guide to eukaryotic genome annotation. Nat Rev Genet 13:329–342. https://doi.org/10.1038/nrg3174

Yang J, Moeinzadeh MHH, Kuhl H, Helmuth J, Xiao P, Haas S, Liu G, Zheng J, Sun Z, Fan W, Deng G, Wang H, Hu F, Zhao S, Fernie AR, Boerno S, Timmermann B, Zhang P, Vingron M (2017) Haplotype-resolved sweet potato genome traces back its hexaploidization history. Nat Plants 3:696–703. https://doi.org/10.1038/s41477-017-0002-z

Yencho C, Pecota KV, Schultheis JR (2008) 'Covington' Sweetpotato. HortScience 43:1911–1914

Zhang L, Wang Q, Liu Q, Wang Q (2009) Sweetpotato in China. In: Loebenstein G, Thottappilly G (eds) The sweetpotato. Springer, Berlin, pp 325–358

Zinn H (2015) A people's history of the United States. A people's history of the United States. HarperCollins, New York

Chapter 4
Spatial and Compositional Diversity in the Microbiota of Harvested Fruits: What Can It Tell Us About Biological Control of Postharvest Diseases

Michael Wisniewski, Samir Droby, Ahmed Abdelfattah, Jia Liu, Susan Whitehead, Shiri Freilich, and Christopher Dardick

Abstract Accumulating evidence indicates that the composition of the microbiota inhabiting an organism (both endo- and epiphytically) can have a profound effect on host physiology and defense responses. The role of the microbiota in plant health and physiology, however, is poorly understood, and few studies have focused on temperate fruit trees. A global effort is underway to characterize the endophytic and epiphytic microbiome of apple fruit with the goal of developing a microbial consortium for the management of a wide range of postharvest diseases, and potentially physiological disorders. Spatial studies of the microbiota of apple peel, calyx-end, stem-end, and pulp tissues have revealed distinct compositional differences, as well as differences between organic and conventional management systems. Other studies have indicated a distinct genotype effect on defining the endophytic microbiota in apple shoots. These data suggest that apple pedigree influences the composition of the endophytic microbiota and that the host and microbiome have co-evolved to

M. Wisniewski (✉) · C. Dardick
U.S. Department of Agriculture – Agricultural Research Service, Kearneysville, WV, USA
e-mail: Michael.Wisniewski@ars.usda.gov

S. Droby (✉)
The Volcani Center, Agricultural Research Organization, Rishon LeZion, Israel
e-mail: samird@volcani.agri.gov.il

A. Abdelfattah
Institute of Environmental Biotechnology, Graz University of Technology, Graz, Austria

J. Liu
Chongqing Key Laboratory of Economic Plant Biotechnology, College of Landscape Architecture and Life Sciences, Chongqing University of Arts and Sciences, Chongquing, China

S. Whitehead
Department of Biological Sciences, Virginia Tech, Blacksburg, VA, USA

S. Freilich
ARO, Newe-Ya'ar Research Center, Ramat Yishay, Israel

D. Spadaro et al. (eds.), *Postharvest Pathology*, Plant Pathology in the 21st Century 11, https://doi.org/10.1007/978-3-030-56530-5_4

some extent, as suggested in the holobiont concept. The implications of greater knowledge of the apple microbiome on disease and cultural management strategies, cultivar breeding, and abiotic stress resistance, are discussed.

Keywords Apple · Endophytes · Epiphytes · Fruit microbiome · Genotype · Holobiont

Introduction

Global estimates of postharvest losses are in the range of 29–38% (FAO 2011), with a value exceeding 50 billion US dollars annually. Control measures rely primarily on synthetic, chemical fungicides but increasingly greater restrictions are being imposed on their use by regulatory agencies due to health and environmental concerns. Finding effective alternative strategies, such as biological control, based on the use of microorganisms naturally occurring on fruit surfaces, has been the focus of considerable research for the last 30 years (Droby et al. 2016).

Although several registered postharvest biocontrol products, based on a single antagonist (mainly yeasts), have been developed, these products have never been used on a large scale due to inconsistent performance under commercial conditions (Wisniewski et al. 2016). The need for alternatives, however, is still valid and the outlook for microbial biocontrol products overall is still very promising. Wisniewski et al. (2016) and Droby et al. (2016) reviewed recent advances in the science of postharvest biological control and other alternative disease control methods. They indicated that the interest in biological approaches to postharvest disease control, and biologically-based products in general, continues to grow, and discussed the drivers underlying the continued interest in biocontrol products. This approach neglects the fact that the introduced antagonist is not the only player in the system, and in general neglects the interactions the antagonist experiences as part of a microbial network and as a component of a biological system (Droby et al. 2016). The field of postharvest biocontrol has reached a crossroads. Existing approaches need to be seriously evaluated and new paradigms and avenues of research need to be considered (Droby et al. 2016). The current paradigm is based on the premise that a single microbial antagonist, isolated from one commodity, can be effectively used to control postharvest decay on different commodities that vary in their genetic background, physiology, postharvest handling, and susceptibility to decay organisms. This paradigm may no longer be appropriate given our current knowledge of microbial ecology, which has been obtained through a growing number of metagenomic studies (Wisniewski et al. 2016; Turner et al. 2013; van der Heijden and Hartmann 2016). A significant gap still exists between basic research underlying the selection of biocontrol agents and their use as commercially-successful products.

The fruit microbiome is a dynamic system and microbe-microbe interactions, as well as environmental conditions, can have a direct effect on community structure and the impact of the community on its host (fruit). Knowledge of the microbiota of harvested fruits and a functional understanding of the impact of community structure on disease resistance and susceptibility is extremely limited. Yet, rhizosphere and phyllosphere studies have clearly demonstrated that microbial community structure can have both a positive and a negative impact on host-related parameters, such as abiotic stress tolerance, nutrient acquisition, hormone signaling, and disease resistance. Dominant microbes have a significant level of network interconnectivity and greatly influence community structure. Understanding microbial networks and their corresponding metabolome would help to identify potential microbe-microbe interactions and the gene products they produce. A significant need exists to develop methods to identify, quantify, and elucidate the metabolic networks constructed by microbial populations that exist on harvested fruit.

One aspect of microbial ecology that has the potential to revolutionize our understanding of plant pathology and biology, is the recent technological advances that have been made in the use of next-generation-sequencing (HTS) and meta-omic technologies to characterize the diversity and function of the microbial communities present in and on host tissues (Abdelfattah et al. 2017). HTS technologies are opening a new frontier in exploring the role of microbes in our environment and their interaction with various hosts (Berg et al. 2016). As presented in this review, microbiome-based studies have the potential of providing information that could cause a fundamental paradigm shift in the way we think about biocontrol strategies, biocontrol products, and postharvest biology in general.

Spatial Differences in the Microbiome of Harvested Apples

Several studies have documented the microbial diversity present on a variety of fruit species, including apple (reviewed by Abdelfattah et al. 2016). These studies have examined the impact of management practices on the composition of the microflora, temporal changes in the microbiome over the course of a growing season, and the composition of the microbiota after harvesting, processing, shipping, and distribution to a local supermarket. Studies documenting the effect of the application of biocontrol agents on the resident microflora have been comprehensively reviewed by Massart et al. (2015), even if most of the reported studies were not related to postharvest biocontrol.

The domesticated apple (*Malus pumila* Mill, family Rosaceae, tribe Pyreae) is a major temperate fruit crop, and is grown in large numbers worldwide. Similar to other fruit crops, apple is subject to infections by several different phytopathogens and is colonized by a number of different microorganisms. Current knowledge about the apple microbiota is largely focused on species of microbes that cause diseases and thus pose economical threats, or, in some cases, natural antagonists that could be used as biological control agents against these pathogens (Wisniewski and Droby 2019). Our current understanding of the fruit microbiota is largely based on the ability to isolate, culture, and identify the most abundant microorganisms, which are now believed to represent only a small fraction of the total estimated microbial diversity (Abdelfattah et al. 2017).

Regarding apple, Abdelfattah et al. (2016) utilizing amplicon-based sequencing demonstrated the presence of a much higher level of diversity of fungi than has ever been reported for apple fruit surfaces. They identified 8000 operational taxonomic units (OTUs), five phyla, and more than 350 genera (Fig. 4.1). Wasserman et al. (2019) similarly documented a high level of bacterial diversity on harvested apples and estimated the presence of over 100 million bacterial cells in one apple. Interestingly, the apple pulp and seeds had the highest numbers of bacterial cells.

In their study, Abdelfattah et al. (2016) indicated that alpha and beta diversity of the fungal microflora of harvested apples differed significantly on different fruit parts. PCoA analysis indicated that the microbiota of samples clustered distinctly based on the part of the fruit from which they were obtained, such as peel, pulp, calyx-end, or stem-end tissues (Fig. 4.2). This strongly indicates that the microflora

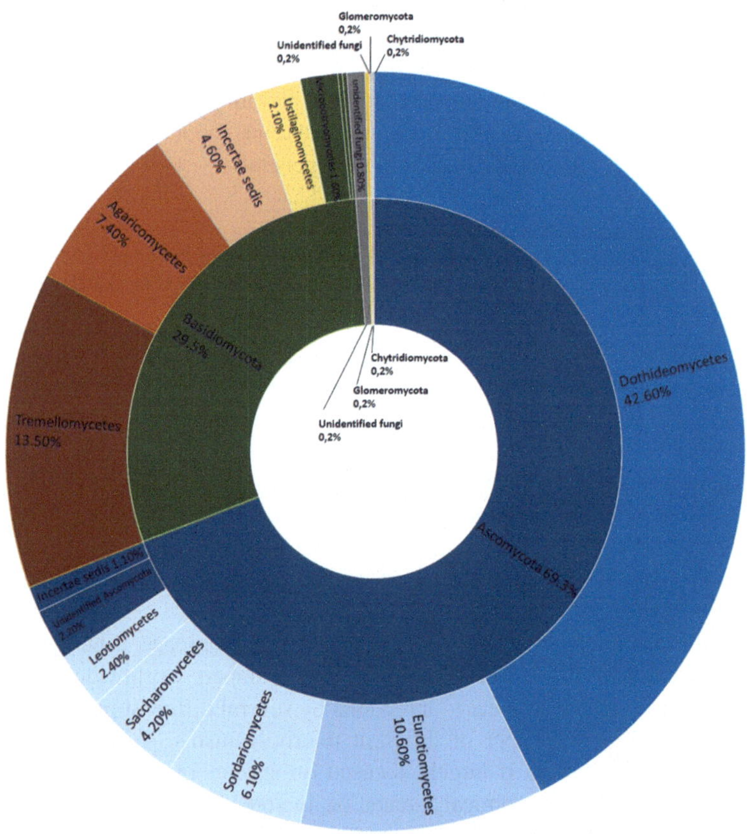

Fig. 4.1 Sunburst chart indicating the relative abundance of fungal phyla and genera identified on the carposphere of Red Delicious apple fruit. (See Abdelfattah et al. (2016) for details on the identification of fungal genera)

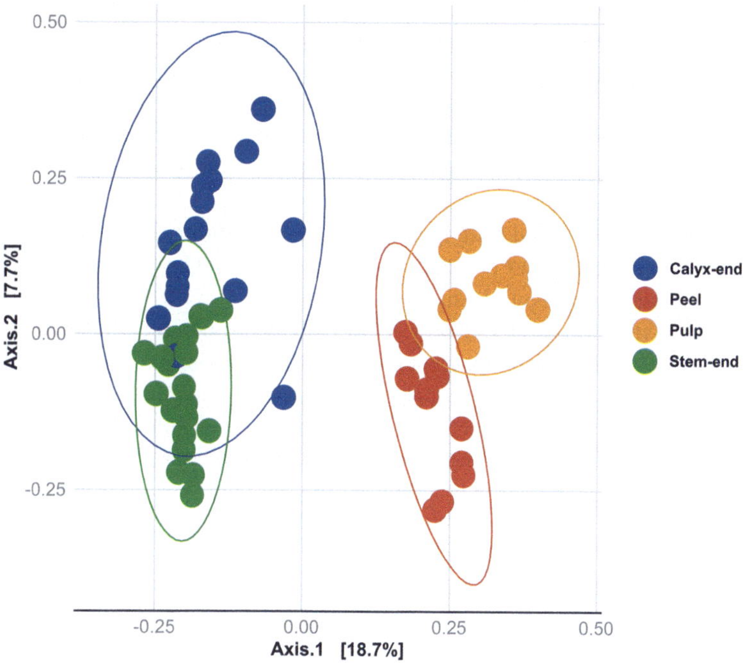

Fig. 4.2 PCoA plot of beta diversity in different portions of Red Delicious apple fruit based on Bray-Curtis dissimilarity metrics. (See Abdelfattah et al. (2016) for details on the identification of fungal genera

associated with different portions of the apple fruit need to be considered when designing biocontrol systems for the management of postharvest diseases. In this study, *Penicillium* was dominant in peel samples, while *Alternaria* was dominant in the calyx- and stem-end samples. *Ascomycota* accounted for over 90% of the observed species, followed by *Basidiomycota* (8%), and *Chytridiomycota* (0.1%). No significant temporal changes were observed in the microbiome over a 2-week storage period at room temperature. Additional work is ongoing to develop spatially-explicit maps of the fruit-associated microbiome and determine how this fine-scale variation relates to fruit metabolome and other aspects of fruit quality.

Another important aspect of the apple microbiome to understand is the determination of a "core microbiome". Is there a specific set of microbial species that inhabit apple fruit regardless of genotype, environment, or geographical location? To answer this question, a global analysis of the apple microbiome of 'Royal Gala' apple at harvest maturity is being conducted. The study involves the participation of researchers from numerous locations in the northern and southern hemisphere, that represent different environmental conditions. Peel, stem-end, and calyx-end samples are being separately collected and analyzed from at least two different orchard sites in each country. The countries include United States (West Virginia, New York, and Washington State), Canada (Ontario and New Brunswick), Uruguay, Israel, Turkey, Italy, Spain, and Switzerland. Such information is essential to determine the evolution of the apple microbiome and how biological strategies can be developed for the management of both pre- and postharvest diseases and physiological disorders of apple. Similar global studies have been conducted on the citrus rhizosphere microbiome (Xu et al. 2018) and the carposphere of nine varieties of wine grapes (Zhang et al. 2019).

In addition to these global efforts, further ongoing studies in the eastern United States are working to identify the major pre-harvest factors that shape variation in the microbiome of harvested fruit, including abiotic conditions, soil properties, pest and pathogen incidence, and pest management practices. Current projects include observational studies of natural variation in commercial orchards as well as experimental studies in research orchards to definitively determine the effects of specific practices on the apple microbiome composition. Considering the well-documented context-dependency of both plant-microbe interactions and biocontrol efficacy in agroecosystems (e.g. Hoeksema et al. 2010; Tylianakis and Romo 2010; Doan et al. 2019), an improved understanding of how multiple interacting factors shape variation in microbial community composition is critical to the development of new biocontrol strategies. It is quite possible that the relative efficacy of various potential postharvest biocontrol products will depend on the fruit microbiome composition and therefore the pre-harvest context in which the fruit was grown.

Genotype Effect on the Composition of the Endophytic Microbiome of Apple Scions and Rootstocks

In a recent study, Liu et al. (2018) examined the effect of different rootstocks (M.9 and M.M.111) on the endophytic microbiome of scion wood of different cultivars: 'Royal Gala', 'Golden Delicious', and 'Honeycrisp'. It was of interest to determine if differences exist in the composition of the microbiota of different apple rootstocks and apple varieties, as well as determine if different rootstock/scion combinations influence the microbial composition of their respective components. Results indicated that *Ascomycota* (47.8%), *Zygomycota* (31.1%), and *Basidiomycota* (11.6%) were the dominant fungal phyla across all samples while the majority of bacterial sequences were assigned to *Proteobacteria* (58.4%), *Firmicutes* (23.8%), *Actinobacteria* (7.7%), *Bacteroidetes* (2%), and *Fusobacteria* (0.4%). The metabarcoding analysis revealed that there was greater fungal diversity than bacterial diversity in both rootstocks and scions. Notably, pedigree had a significant effect on the composition of the endophytic fungal microbiota of scions where closely-related cultivars ('Golden Delicious' and 'Royal Gala') exhibited a microbial community that was more similar to each other than they were to more distantly-related scion cultivars (Fig. 4.3). A similar relationship was not observed in bacterial taxa. Interestingly, 'M.M.111', which is classified as a vigorous rootstock, was richer in growth-promoting bacteria, compared to 'M.9', which is classified as a dwarfing rootstock. Additional projects are underway to assess how the fruit microbiome

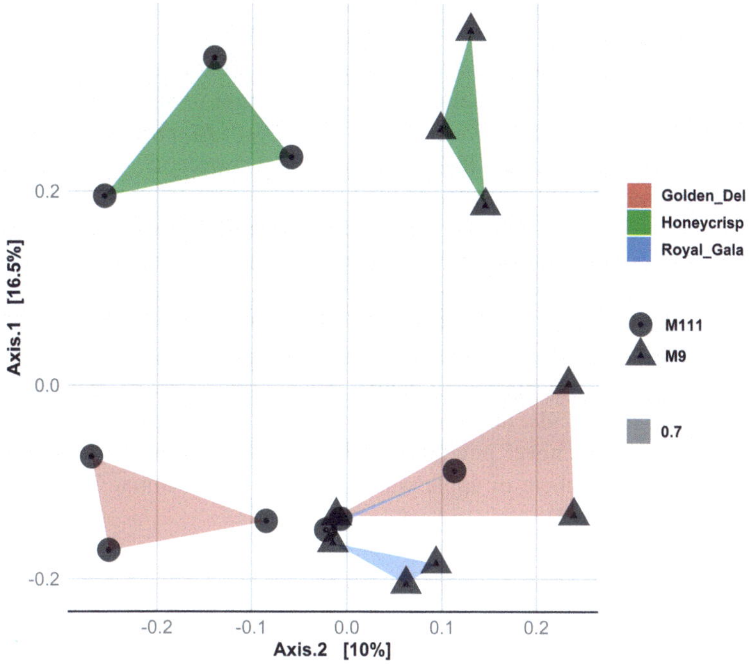

Fig. 4.3 PCoA plot of beta diversity of the endophytic microbiome of scions of different cultivars of apple (Honeycrisp, Golden Delicious, and Royal Gala) grafted on two different rootstocks (M.9 and M.M.111) based on Bray-Curtis dissimilarity metrics. (See Liu et al. (2018) for details)

varies across eight cultivars, and how cultivar may interact with other management practices to shape the fruit microbiome. However, considering the numerous potential genotype-environment interactions that likely play a role in shaping the fruit microbiome, much work is needed in this area. Although the mechanisms by which the plant genotype, either rootstock or scion, have a determinant effect on the composition of microbial community is not yet known, our current knowledge conforms to the "holobiont" concept in which metaorganisms (plants and their associated microbes) have co-evolved.

The Apple Rhizosphere

The rhizosphere component of the plant microbiome has been the component that has been most studied, as is evident from the number of publications about the rhizosphere microbiome compared to studies of the carposphere one. While not the subject of the present review, it should be noted that the composition of the rhizosphere microbiome is considered to have a major impact on the growth, vigor, and stress and disease resistance of a plant. Indeed, rhizosphere research has been the driver behind the recent development of several new 'biological' products and whole new companies.

In this regard, Mazzola and colleagues, in a series of studies, have extensively studied the rhizosphere community of apple in relation to apple replant disease (Wang and Mazzola 2019; Hewavitharana et al. 2019; Mazzola et al. 2014). In their work, these researchers have examined the impact of various types of "Brassicaceae-based" soil amendments and anaerobic soil disinfestation methods on both nematode populations and the apple rhizosphere microbiome. In addition to studying the composition of the rhizosphere and surrounding soil microbiome, they have also characterized the impact of various management practices on the community metabolome (Hewavitharana et al. 2019). Their results clearly indicate that various management strategies affect "community metabolism" by temporally altering specific classes of labile compounds, such as methyl sulfide compounds, organic acids, and p-cresol with antimicrobial properties, and that these alterations offer novel potential modes of disease control. A community analysis of the rhizosphere of apple was conducted in the Bohai Gulf region of China (Jiang et al. 2017).

Conclusion

Understanding the functioning of epiphytic, endophytic, and rhizosphere microbial communities and their influence on host physiology and disease resistance is a new frontier that will modify many of the existing paradigms that exist about biocontrol systems, stress tolerance, and plant growth and development. Wisniewski and Droby (2019) have discussed the importance of studying the fruit microbiome and noted that while the challenges posed by this new frontier are many, the opportunities for addressing many agricultural problems are great, including the development of function-based biocontrol systems and biological-based strategies for addressing postharvest physiological disorders. Indeed, they provided a model indicating the interactions that occur between fruit hosts, the natural fruit microbiota, and fruit pathogens (Fig. 4.4).

Regarding postharvest diseases and biocontrol systems, significant foundational questions regarding the characterization of core microbiomes on different commodities, and the impact of management, harvesting, processing, and storage systems on the functioning of native microbial communities, still need to be addressed. It is not unreasonable to predict that increased knowledge of microbial community systems will lead to the development of synthetic consortia that can be used to

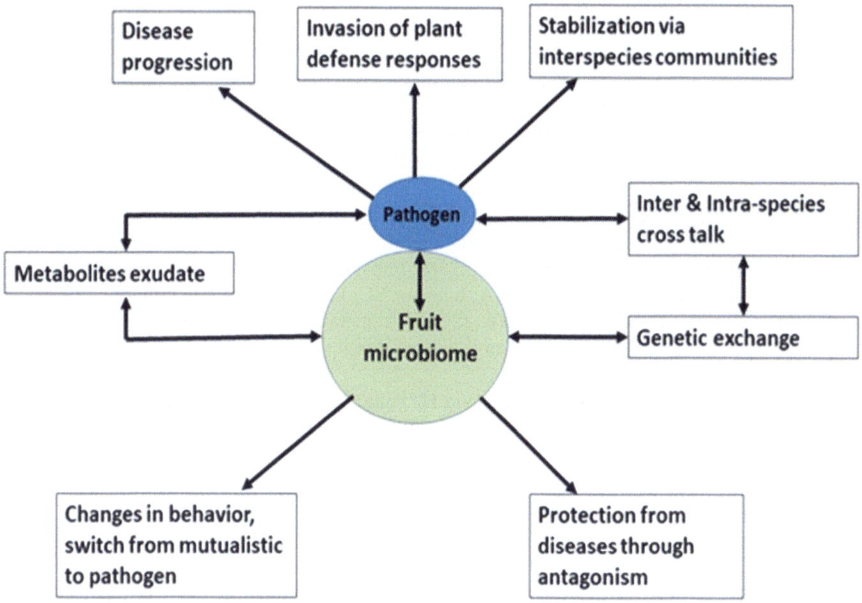

Fig. 4.4 Schematic model illustrating the comprehensive interactions that occur between the various components and factors present on and in the apple carposphere. (See Wisniewski and Droby (2019) for details)

prevent postharvest diseases and mitigate physiological disorders in harvested commodities.

Acknowledgements Partial funding for the described completed and ongoing projects have been provided through a grant from the U.S. – Israel Binational Agricultural Fund, (IS-5040-17), an Agriculture and Food Research Initiative Competitive Grant no. 2018-07366 from the USDA National Institute of Food and Agriculture, and the European Union's Horizon 2020 under "Nurturing excellence by means of cross-border and cross-sector mobility" program for MSCA-IF-2018-Individual Fellowships, grant agreement 844114.

References

Abdelfattah A, Wisniewski M, Droby S, Schena L (2016) Spatial and compositional variation in the fungal communities of organic and conventionally grown apple fruit at the consumer point-of-purchase. Hortic Res 3:16047. https://doi.org/10.1038/hortres.2016.47

Abdelfattah A, Malacrino A, Wisniewski M, Cacciola SO, Schena L (2017) Metabarcoding: a powerful tool to investigate microbial communities and shape future plant protection strategies. Biol Control 120:1–10. https://doi.org/10.1016/j.biocontrol.2017.07.009

Berg G, Rybakova D, Grube M, Köberl M (2016) The plant microbiome explored: implications for experimental botany. J Exp Bot 67:995–1002. https://doi.org/10.1093/jxb/erv466

Doan HK, Maharaj NN, Kelly KN, Miyao EM, Davis RM, Leveau JH (2019) Antimycotal activity of Collimonas isolates and synergy-based biological control of Fusarium wilt of tomato. Phytobiomes J 4:64–74. https://doi.org/10.1094/PBIOMES-05-19-0027-R

Droby S, Wisniewski M, Teixidó N, Spadaro D, Jijakli MH (2016) The science, development, and commercialization of postharvest biocontrol products. Postharvest Biol Technol 122:22–29. https://doi.org/10.1016/j.postharvbio.2016.04.006

FAO report (2011) Global food losses and food waste: extent, causes and prevention. FAO, 730 Rome. http://www.fao.org/docrep/014/mb060e/mb060e.pdf

Hewavitharana SS, Klarer E, Reed AJ, Leisso R, Poirier B, Honaas L, Rudell DR, Mazzola M (2019) Temporal dynamics of the soil metabolome and microbiome during simulated anaerobic soil disinfestation. Front Microbiol 10:2365. https://doi.org/10.3389/fmicb.2019.02365

Hoeksema JD, Chaudhary VB, Gehring CA, Johnson NC, Karst J, Koide RT, Pringle A, Zabinski C, Bever JD, Moore JC, Wilson GW (2010) A meta-analysis of context-dependency in plant response to inoculation with mycorrhizal fungi. Ecol Lett 13:394–407. https://doi.org/10.1111/j.1461-0248.2009.01430.x

Jiang J, Sung Z, Yeng X, Mao Z, Nie X, Guo H, Peng X (2017) Microbial community analysis of the apple rhizosphere around Bohai Gulf. Sci Rep 7:8918. https://doi.org/10.1038/s41598-017-08398-9

Liu J, Abdelfattah A, Norelli J, Burchard E, Schena L, Droby S, Wisniewski M (2018) Apple endophytic microbiota of different rootstock/scion combinations suggests a genotype-specific influence. Microbiome 6:18. https://doi.org/10.1186/s40168-018-0403-x

Massart S, Perazzolli M, Höfte M, Pertot I, Jijakli MH (2015) Impact of the omic technologies for understanding the modes of action of biological control agents against plant pathogens. BioControl 60:725–746. https://doi.org/10.1007/s10526-015-9686-z

Mazzola M, Hewavitharana SS, Strauss SL (2014) Brassica seed meal soil amendments transform the rhizosphere microbiome and improve apple production through resistance to pathogen re-infestation. Phytopathology 105:460–469. https://doi.org/10.1094/PHYTO-09-14-0247-R

Turner TR, James EK, Poole PS (2013) The plant microbiome. Genome Biol 14:209. https://doi.org/10.1186/gb-2013-14-6-209

Tylianakis JM, Romo CM (2010) Natural enemy diversity and biological control: making sense of the context-dependency. Basic Appl Ecol 11:657–668. https://doi.org/10.1016/j.baae.2010.08.005

Van Der Heijden MGA, Hartmann M (2016) Networking in the plant microbiome. PLoS Biol 14:e1002378. https://doi.org/10.1371/journal.pbio.1002378

Wang L, Mazzola M (2019) Field evaluation of reduced rate Brassicaceae seed meal amendment and rootstock genotype on the microbiome and control of apple replant disease. Phytopathology 189:1378–1391. https://doi.org/10.1094/PHYTO-02-19-0045-R

Wasserman B, Müller H, Berg G (2019) An apple a day: which bacteria do we eat with organic and conventional apples. Front Microbiol 10:1629. https://doi.org/10.3389/fmicb.2019.01629

Wisniewski M, Droby S (2019) The postharvest microbiome: the other half of sustainability. Biol Control 137:104025. https://doi.org/10.1016/j.biocontrol.2019.104025

Wisniewski M, Droby S, Norelli J, Liu J, Schena L (2016) Alternative management technologies for postharvest disease control: the journey from simplicity to complexity. Postharvest Biol Technol 122:3–10. https://doi.org/10.1016/j.postharvbio.2016.05.012

Xu J, Zhang Y, Zhang P, Trivedi P, Riera N, Wang Y, Liu X, Fan G, Tang J, Coletta-Filho HD, Cubero J, Deng X, Ancona V, Lu Z, Zhong B, Rober MC, Capote N, Catar V, Pietersen G, Verniere C, Al-Sadi A, Li L, Yang F, Xu X, Wang J, Yang H, Jin T, Wang N (2018) The structure and function of the global citrus rhizosphere microbiome. Nat Commun 9:4894. https://doi.org/10.1038/s41467-018-07343-2

Zhang S, Chen X, Zhong Q, Zhuang X, Bai Z (2019) Microbial community analyses associated with nine varieties of wind grape carposphere based on high-throughput sequencing. Microorganisms 7:668. https://doi.org/10.3390/microorganisms7120688

Chapter 5
Endophytic Microbiome in the Carposphere and Its Importance in Fruit Physiology and Pathology

Ajay Kumar, Yeka Zhimo, Antonio Biasi, Shoshana Salim, Oleg Feygenberg, Michael Wisniewski, and Samir Droby

Abstract Each plant has a microbiome that is consisted of epiphytic and endophytic microbial communities. The plant microbiome may play a prominent role in different functions such as growth, disease, suppressing pathogens and potential influence of the physiology of reproductive plant organs (e.g., fruit). Endophytic microbiomes of any plant organs have mutualistic interaction between each other and functionally interact with each other in multiple ways. The distribution pattern of endophytic microbiota varies with each plant's organs and largely depends on plant genotype (i.e., cultivar) developmental stages, growth condition, biotic and abiotic factors. The distribution pattern, as well as the potential utilization of endophytic microbiota that are associated with internal tissues for managing pre and postharvest pathogens, is discussed in this chapter.

Keywords Biotic stress · Biocontrol agents · Carpophore · Endophytes · Endophytic diversity

---

A. Kumar · Y. Zhimo · A. Biasi · S. Salim · O. Feygenberg · S. Droby (✉)
Department of Postharvest Science, ARO, The Volcani Center, Rishon Lezion, Israel
e-mail: samird@volcani.agri.gov.il

M. Wisniewski
Appalachian Fruit Research Station, U.S. Department of Agriculture – Agricultural Research Service, Kearneysville, WV, USA

D. Spadaro et al. (eds.), *Postharvest Pathology*, Plant Pathology in the 21st Century 11, https://doi.org/10.1007/978-3-030-56530-5_5

Introduction

The carposphere of agricultural crops harbors diverse microbial communities as epiphytes or endophytes that play an integral role in the biology, physiology and resistance of fruits and vegetables to pre and postharvest decay (Droby and Wisniewski 2018). The composition and function of both epiphytic and endophytic microbial populations in the carposphere, however, is still unexplored in relation of their involvement in fruit physiology, quality, and management of pathogen attack. Extensive research work has been published regarding the use of epiphytic yeasts and bacteria as biocontrol agents of postharvest diseases (Droby et al. 2019), but a very limited number of reports is available on microbial endophytes of fruit and their potential functions and utilization in a beneficial manner against biotic and abiotic stressors.

The term "endophyte" was used first for the fungi that were isolated from the internal cells/tissues of host plants, but later the concept had been changed, and bacterial communities were also considered as endophytes (Chanway 1996; Hardoim et al. 2015). De Bary (1866) was the first to define microorganisms growing inside plant tissues as endophytes. Later on, the definition, as well as types of endophytes, were modified as per researcher observations. Hallmann et al. (1997) defined endophyte as the microbes that can be isolated from surface-disinfected plant tissues, and those microbes that could survive inside their host system without causing any disease. Further, Petrini (1991) defined endophytes as the organisms inhabiting plant tissue at some times in their life cycle without causing any infection to the host plant. Some researchers have also categorized endophytes on the basis of their types, i.e., bacteria or fungi, and their relationship with the plants such as facultative or obligate (Rosenblueth and Martínez-Romero 2006). Hardoim et al. (2015), however, have characterized endophytes on the basis of colonization niche instead of their function.

In this chapter, we tried to summarize the work that has been done on endophytes in the carposphere and describe the diversity, distribution pattern as well as the potential utilization of endophytic microbiota that are associated with internal tissues for managing pre and postharvest pathogens.

Entry, Colonization and Transmission of Endophytes into Plant Tissue

In general, the entry and colonization of host plant tissue by microbes is a complex process and is controlled by various plant-microbe interactions mediated by signaling molecules and secretory products. Attachment of the microbial cells to the host surface is the initial step of endophytic colonization. The production and secretion of various products such as exopolysaccharides (EPS), lipopolysaccharides (LPS) and cell surface saccharides were demonstrated to be involved in adhesion with the host surface. The appendages like flagella, fimbriae and pili in bacteria facilitate chemotactic movement towards the host tissue and start colonization (Sauer and Camper 2001; Zheng et al. 2015).

The most preferred site for microbial entrance are the aerial parts of the plants, root zone, wounds, cotyledons, and stomata/lenticels. After entry, microbes move inside the tissue by passive or an active process (Zinniel et al. 2002). The microbial entry through the natural opening like stomata, hydathodes, is usually facilitated by passive movement. However, penetration using secretory products such as enzymes, EPS, LPS, and secondary metabolites could be considered as an active process for endophytic colonization (Böhm et al. 2007; Suárez-Moreno et al. 2010).

The interaction of endophytic microorganisms with the host during penetration of plant tissue is quite different from that of pathogenic strains. Generally, pathogenic microorganisms secrete a large amount of cell wall degrading enzymes in and around the infection site that actively degrade the tissue and facilitate easy access for the pathogen. However, endophytes secrete very small quantities of cell wall degraders, such as cellulase and pectinase, that are notable to trigger the immune response of host plants (Elbeltagy et al. 2000; Reinhold-Hurek et al. 2006). Successful colonization of endophytes involves cross-talk between the host through signaling molecules (Compant et al. 2010; Brader et al. 2014).

Signaling molecules were reported to mediate communication and cooperative behaviors between endophytes and the host plant (Keller and Surette 2006). Some of the genes, including those involved in nitrogen fixation and metabolism (ntrY and ntrX), as well as cell fate control (pleC and pleD), are found more predominantly in the endophytes in comparison to rhizobacteria and phytopathogens (Hardoim et al. 2015). Numerous biological processes related with energy regulation and utilization, such as carbon and nitrogen fixation, electron transport, aerobic and anaerobic respiration, are known to be regulated in response to cellular redox balance, that helps in endophyte colonization inside the host plant (Elsen et al. 2004; Hardoim et al. 2015).

Additional factors take part in successful endophytic colonization, including genotypes of the host, specificity of microbial strains, and nutrients availability (Hardoim et al. 2015). Compant and his co-workers (2008) studied in detail the colonization route of *Paraburkholderia phytofirmans* PsJN and reported that its entry into the plant tissue is made through the exodermis layer and cortical cells of the roots and tackle the barrier of endodermal layer leading its way to the central zone. From the central zone, endophytes spread towards the upper part of the plant through the xylem vessels. Generally, endophytes that enter through the root system prefer un-suberized endodermal cells of the apical roots (James et al. 2002; Gasser et al. 2011). Once microbes are endophytically borne, they utilize the nutrients available in the host plant tissue (Rasche et al. 2009) as a carbon source for survival and proliferation (Malfanova et al. 2013). The source of energy, however, varies with microbial strains. For example, an endophytic strain of *pseudomonads* uses L-arabinose as a source of energy in cucumber plants (Iwai et al. 2003). Diazotrophic endophytic isolates of rice *Enterobacter dissolvens*, *Brevundimonas aurantiaca*, *Pantoea agglomerans*, *Pseudomonas spp.* and Enterobacteriaceae used L-arabinose as well as glucose (Prakamhang et al. 2009) whereas Methylobacterium utilizes methanol as a source of energy (Sy et al. 2005).

The transmission of microbes to the host may occur horizontally (through environments), vertically (within the plant, most commonly through seed) or mixed (Bright and Bulgheresi 2010; Frank et al. 2017). However, the diversity of bacterial community inside the seeds grown under aseptic conditions is comparatively lower than in environmental grown plants, suggesting that the horizontal mode is the preferred transmission mode of endophytic microorganisms (Hardoim et al. 2015; Compant et al. 2005). However, it is possible that some endophytes are transmitted both vertically and horizontally or mixed type. Endophytes that are consistently transferred across generations must have a route from seeds to reproductive organs (fruits), either via xylem vessels or via the shoot apical meristem that differentiates into reproductive organs (Frank et al. 2017).

Coevolution of Endophytes with Their Host

The genotype of the plant may influence the composition of the endophytic microbiome. In this relation, Liu et al. (2018) demonstrated a genotype-specific influence on the apple endophytic microbiota of three apple cultivars: Royal Gala, Golden Delicious and Honeycrisp, rootstock/scion combination. They observed differences in the fungal endophytic microbiota of all the three cultivars. However, no difference was observed in the endophyte bacterial communities. The composition of fungal diversity varies among all the cultivars; however, diversity in "Golden Delicious" and "Royal Gala" was not significantly different between each other's, but they were distinctly separated from "Honey Crisp" regardless of the genotype of the rootstock. Furthermore, the pedigree analysis of all three cultivars showed Royal Gala and Golden delicious close to each other and distant from Honey Crisp.

The intimate association of endophytic microbes with plant organs describes the ecological and evolutionary relationship between each other. The life span expended by endophytes inside the host plant, the ability to grow within the host tissue and the age of sexual maturity of the plants greatly constrain the microbial growth pattern within the host plants (Saikkonen et al. 2004). The annual plants complete their life cycle growth and reproduction in one season; however, perennial hosts might be able to nurture and cultivate the endophytic microbiome in a better way in comparison to the annual host.

Diversity of Endophytic Microbiota on Fruit Carpophore

Compant et al. (2011) studied the colonization pattern of endophytic bacterial strains inside flowers, berries and seeds of grapevine plants and compared that with different plant parts. Endophytes were detected in almost all the plant parts, but their densities vary according to plant organs. The abundance of microbiota was lower in the flowers, berries and seeds in comparison to other parts of the plant (e.g., leaves, stems). However, regarding the reproductive parts, higher diversity was detected in the flowers followed by pulp and lowest in the seeds of the grape berry.

The abundance and diversity of endophytic microbiota in the carpophore also depend on the developmental stages of the fruit. In this regard, Miguel et al. (2013) reported a diversity pattern of endophytic bacteria in *Coffea canephora* at three developmental stages: green, green-yellow and ripe. They observed that α- and γ-Proteobacteria, Actinobacteria, Firmicutes, and Bacteroidetes were the most prevalent community. *Kocuria turfanensis* and *Pantoea vagans, Bacillus thuringiensis, Bacillus licheniformis, Agrobacterium tumefaciens, Escherichia coli, Enterobacter hormaechei, Chryseobacterium* sp., *Ochrobactrum* sp. were the most common species, and the diversity of the bacterial strains varies according to the developmental stage of the fruit. Higher diversity was observed highest at the green stage and lowest in the ripe stage. Among bacterial species, *Bacillus subtilis* was predominantly present during the early stages of development, whereas *Klebsiella oxytoca* was dominant in the ripe stage. It was suggested that a higher concentration of caffeine and sugar at the ripe stage might be related to lower diversity.

There was an effect of the ripening stage of the fruit on endophytic fungal diversity of the sea buckthorn (*Hippophae rhamnoides* L) (Lukša et al. 2020). They observed that both unripe and ripe stages had different fungal communities; unripe berries were shown to be colonized by Aureobasidium, Taphrina and Cladosporium, while ripe berries were dominated by the yeast genera Aureobasidium and Metschnikowia.

In recent work, Cruz et al. (2019) reported that cultural practices affected endophytic microbial diversity in peels of five tropical fruits (banana, guava, mango, papaya, and passion fruit) grown in the wild and in commercial orchards. They observed higher diversity of bacterial communities in the peel of all five fruits compared to fungal communities, irrespective of the production system. Furthermore, fruit type and production practices had a significant effect on the abundance of classes of endophytic microorganisms. OTU number in the wild papaya sample was relatively high in comparison to the conventional sample. No significant differences were observed in endophytic bacterial community diversity in peel samples of conventional and wild cultivated fruit. In mangoes, however, fungal OTUs were highest in peels of conventionally produced fruit independent of the cultivation practice. The fungal diversity in both the conventional and non-cultivated samples was comparable in all the fruit types except in mango and guava, where a significant difference was observed for conventionally produced and wild grown fruit peel samples.

In watermelon, Saminathan et al. (2018) studied the impact of red (PI459074, Congo, and SDRose) and yellow (PI227202, PI435990, and JBush) fruit-flesh cultivars, grown in different geographic regions, season, and fruit ripening stage on the endophytic microbiota. The metabarcoding studies showed that each cultivar possesses specific microbiota, similar to the results of previously published reports (Frank et al. 2017; Hardoim et al. 2015). Proteobacteria were abundantly present in the cultivar SDRose and PI227202, while, Cyanobacteria were dominant in the

Congo and PI4559074 cultivars. They also concluded that the presence of disease in the SDRose cultivar limits microbial diversity.

In another stufy, Glassner et al. (2015) investigated the endophytic microbiome of cucurbit fruits (Cucumis *melo Reticulatus* group of melon) collected from two different sites, one from the commercial cultivation farm and the other from the wild. In both sites, the most prevalent endophytic bacterial class was α-, β-, and Gamma γ-proteobacteria, Firmicutes and Actinobacteria. Visualization and quantification of endophytic bacteria using DOPE-FISH microscopy showed substantial differences between the wild and domesticated cucurbit taxa in regard to the abundance and the diversity of endophytes.

Seasonal variation during the cultivation period also has an effect on the endophytic microbial composition of fruit. In mulberry fruit, endophytic microbial diversity was found to be highly dynamic, and seasonal variation act as a determinant factor in microbial composition inside the host tissue, followed by the effect of genetic backgrounds of the different cultivars. For example, colonization of endophytes generally increased during rainy and warm seasons (Ou et al. 2019).

The diversity of microbial endophytes in the caroposphere was reported to be dependent on the health status of the fruit. In this regard, Wu et al. (2019) studied the endophytic microbial diversity of healthy and diseased kiwifruit. They found that healthy fruits had higher microbial diversity and unique composition compared to the diseased fruit. In diseased fruit, the microbiome was dominated by the Botryosphaeriaceae fungal family, while endophytes in healthy fruit were dominated by the Pleosporaceae family. In the case of the bacterial community, the abundance of Methylobacteriaceae, Sphingomonadaceae, Nocardioidaceae families was higher in the healthy fruit. On the functional level, the endophytic microbiome of healthy fruit had higher relative abundance of ABC transporter genes, two-component system genes related to bacterial secretion systems and chemotaxis, but lower abundance of genes related with degradation of polycyclic aromatic hydrocarbon, amino and nucleotide sugar metabolism, cific aminoacids such as glycine, serine and threonine metabolism.

Al-Kharousi et al. (2016) studied the endophytic community in fresh fruits and vegetables originated from different countries and local sources in Oman and observed that Enterobacteriaceae occurred in 60% of fruits and 91% of vegetables. Enterococcus was isolated from 20% of the fruits and 42% of vegetables. However, *E. coli* and *Sraphylococcus. aureus* were isolated from 22% and 7% of vegetables, respectively. On the species level, *E. coli, Klebsiella pneumoniae*, *Enterococcus casseliflavus*, and *Enterobacter cloacae* were the most abundant species among all the samples.

In Jingbai pear (*Pyrus ussuriensi* Maxim.), the endophytic microbiota in the different plant organs was characterized, showing that each tissue type of the plant had a specific bacterial community (Ren et al. 2019a). The abundances and diversity of bacterial communities were highest in the root tissue followed by flower, stem, and fruit. The lowest abundance and diversity were recorded in the leaves (Ren et al. 2019a, b).

In tomatoes grown under greenhouse condition, endophytes of bacterial communities in roots, stem, leaves, fruit and seeds were investigated by Dong et al. (2019). The findings of this study showed that stem had the highest diversity and abundance of endophytes followed by flowers and fruit. In general, the lowest diversity and abundance were observed in the phyllosphere tissue. Proteobacteria was the most dominant phyla among all the sampled tissue except seeds and fruit jelly of tomatoes. The genera Acinobacter, Enterobacter and Pseudomonas were dominant in the root, stem and leaves tissues, whereas, in the fruit, diversity of endophytes differed in each compartment. The pericarp and seeds of the fruit were dominated by Enterobacter; placenta with Actinobacter, whereas the jelly tissues of the fruit were dominated by Weissella.

The endophytic bacterial communities in the different parts of tomato fruit were also reported by Ottesen et al. (2013). This work revealed that fruit had specific microbial communities varied in different parts of the plant. It was speculated that ovaries, which turn into the flesh of the fruit, create an unique environment that harbors specific microbial communities (Aleklett et al. 2014). The fact that different parts of the same fruit had unique microbiome was also supported by Krishnan et al. (2012), who showed different and specific endophytic bacterial communities in different parts of papaya fruits. Kocuria, Acinetobacter and Enterobacter species were found in the seed as well as endocarp of the fruits, whereas *Staphylococcus sp.* was in the mesocarp and *Bacillus* spp. were predominantly present throughout the different parts of the fruit.

Five fungal strains in fruit flesh and seeds of Jambolana were isolated: *Neofusicoccum parvum*, *Pestalotiopsis* sp., *Phomopsis* sp. and *Colletotrichum fructicola* (Hanin and Fitriasari 2019). In strawberry fruit, Hipol et al. (2014) isolated fungal endophytes and evaluated their antioxidant properties. The endophytic fungi were: *Sporidiobolus* sp., *Rhodotorula* sp., *Pilidium concavum*, *Corynespora cassiicola*, *Neodeightonia subglobosa*, *Aspergillus awamori*, and *Aspergillus* sp. *Aspergillus awamori* DT11 and *Corynespora cassiicola* DT13 had a strong antioxidant potential and it was suggested that these fungal strains may be a possible source of the strawberry fruits nutritive value. In another work, however, De Melo et al. (2012) reported that *Bacillus subtilis*, *Bacillus* sp. *Enterobacter* sp. *Enterobacter ludwigii*, *Lactobacillus plantarum*, *Pseudomonas* sp. *Pantoea punctate*, *Curtobacterium citreum* and *Arthrobacter* sp. constituted the endophytic bacterial communities in strawberry fruits.

A summary of published works about endophytic microbial strains/genera isolated from different parts of the fruits are presented in Table 5.1.

Stem-End Rots Pathogens in Fruit as Endophytes

Besides non-harmfull microbial endophytes, pathogenic strains reside as endophytes in various parts of fruits in the quiescent stage and turn pathogenic when fruit reach a ripening stage in which favorable conditions exist. The stem-end rots (SER)

Table 5.1 Endophytic microorganism isolated from the different parts of the fruit or fruit caropsphere

Fruits	Specific parts of fruits	Genera/Strains	References
Apple	N.M.	*Schwanniomyces vanrijiae, Galactomyces geotrichum, Pichia kudriavzevii, Debaryomyces hansenii*, and *Rhodotorula glutinis*	Madbouly et al. (2020)
Brazilian tropical fruits (Banana, Guava, Mango, Papaya, and Passion fruit)	Peel	Proteobacteria, Actinobacteria, and Bacilli.	Cruz et al. (2019)
Coffee (*Coffea canephora*)	N.M.	*Bacillus subtilis, Klebsiella oxytoca*	Miguel et al. (2013)
Coffee (*Coffea Arabica*)	N.M	Bacillus, Staphylococcus, and Paenibacillus	Oliveira et al. (2013)
Cucurbits	N.M.	*Bacillus subtilis*, *B. licheniformis*, *B. amyloliquefaciens*, *Burkholderia*	Glassner et al. (2015)
Grape berries	N.M.	Metschnikowia, Pichia, Hanseniaspora, Acinetobacter, Burkholderia, Bacillus, Acetobacter, Gluconobacter	Hall and Wilcox (2019)
Grapes	Pulp	*Bacillus cereus*	Compant et al. (2011)
Grape wine	N.M.	Bacteria, Fungi, Yeast	Liu et al. (2018)
Guava	N.M.	*Saccharomycopsis fibuligera*	Abdel-Rahim and Abo-Elyousr (2017)
Jambolana	N.M.	*Neofusicoccumparvum, Pestalotiopsis*	Hanin and Fitriasari (2019)
Noni (*Morinda citrifolia* L.)	N.M.	Paenibacillus	Liu et al. (2018)
Papaya	Seed, Mesocarp, Endocarp	Kocuria, Acinetobacter Enterobacter, Bacillus Staphylococcus	Krishnan et al. (2012)
Strawberry	N.M.	*B. subtilis*, *Enterobacter* sp., and *Pseudomonas* sp.	De Melo et al. (2012)
Strawberry	N.M.	*Sporidiobolus* sp., *Rhodotorula* sp., Pilidium concavum, *Corynespora cassiicola, Neodeightonia subglobosa, Aspergillus awamori*, and Aspergillus sp.	Hipol et al. (2014)
Sweet cherries	Flesh	*Aureobasidium pullulans*	Schena et al. (2003)
Rambutan (*Nephelium lappaceum* L.)	N.M.	Corynebacterium, Bacillus, Chryseobacterium, Staphylococcus	Suhandono et al. (2016)

(continued)

Table 5.1 (continued)

Fruits	Specific parts of fruits	Genera/Strains	References
Star fruit (*Averrhoa carambola*)	N.M.	*Biscogniauxia capnodes*	Sritharan et al. (2019)

N.M. = not mentioned about the specific parts of the fruits

pathogens are major diseases of ripe and ready to eat fruits. The SER causing pathogens generally enter into the stem through the natural openings and wounds, similarly to endophytic strains at the time of inflorescence or flowering (Johnson et al. 1992; Galsurker et al. 2018), and reside asymptomatically as endophytes mainly in the xylem in the stem tissue, but also in the phloem, until fruit ripening (Prusky et al. 2013; Galsurker et al. 2018). The transition from endophytic to necrotrophic lifestyle might be due to biochemical contents of sugar, phytohormones or physiological changes in the ripe fruit, in which a decrease in the level of antifungal compounds takes place along with a change in the surrounding pH (Alkan and Fortes 2015).

Utilization of Endophytic Microorganisms in Diseases Management of Fruits

Currently, plant endophytic microbial organisms are mostly utilized for growth promotion, stress management, and biocontrol agents against a range of pests pathogens in the pre and postharvest phases. Several of these endophytes were shown to have the capability to synthesize a broad range of metabolites, antibiotics, volatile compounds, hormones, exopolysaccharides and cell wall degrading enzymes (Fig. 5.1) (Singh et al. 2017; Nigris et al. 2018; Shameer and Prasad 2018). Additionally, endophytes may also be involved in signaling as well as regulation of metabolic pathways in the fruit via synthesizing enzymes, ACC synthase activity, quorum sensing, etc.…

Generally, most of the antagonistic microorganisms used for the control of postharvest diseases of fruits were isolated from the epiphytic population of the carposphere, phyllosphere or rhizosphere, and only a few were of an endophytic origin. Here we focus on the potential use of endophytic strains in the control of pathogens that are associated with fruits.

In early studies, endophytic microbial strains were suggested by Conway et al. (2007) and Janisiewicz et al. (2013) as antagonistic microorganism against postharvest fruit diseases. The additional value of endophytic microbiota as biocontrol agents is the potential of synthesis of a wide variety of secondary metabolites (including phytohormones), which can directly or indirectly be involved in regulating the physiology, maturation and ripening of the fruit (Lodewyckx et al. 2002;

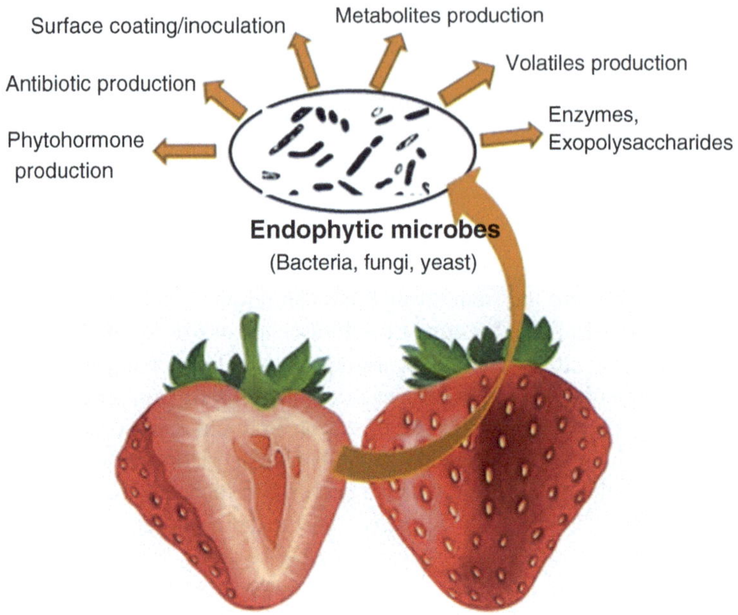

Fig. 5.1 Overview of endophytic microorganism and their action mechanism as antagonistic agent

Suhandono et al. 2016; Singh et al. 2019, 2020). In this relation, endophytes, in comparison to the epiphytic microbiota, could be better colonizers due to better adaptability and acclimatizing nature inside the host tissue (Kumar et al. 2020).

Various reports showed the effect of endophytic microbial strains in the management of various fruit diseases (Hallmann et al. 1997; Sturz and Christie 1996). Istifadah et al. (2017) utilized endophytic strains of *Lysinibacillus* sp., *Bacillus subtilis*, *Azotobacter chrococcum* and *Pseudomonas cepacea* alone or in combination to inhibit the incidence of bacterial wilt disease in chili plant. An endophytic strain of *Pseudomonas putida* biovar A, from the pericarp of papaya, was used to control *Phytophthora nicotianae* infection on harvested fruit (Shi et al. 2010). This bacterial antagonist was proven to have the capability to colonize various compartments of fruit lamina, leafstalk, pericarp and pulp of papaya and also inhibit the growth of several types of pathogens. It was highly effective as well against anthracnose of harvested papaya fruit. Preharvest application of *P. putida* significantly reduced the anthracnose disease index, with the best control effect reaching up to 63% when the application was done immediately after the florescence stage. *Colletotrichum orbiculare,* the causal agent of cucumber anthracnose, was successfully controlled by endophytic actinomycete strains of *Streptomyces* sp. (Shimizu et al. 2009). Lai et al. (2012) utilized the endophytic strain of *Paenibacillus polymyxa*, isolated from root tissue of *Sophora tonkinensis*, for the control of the postharvest disease green mold decay of citrus.

Endophytic lactic acid bacteria (LAB), isolated from papaya seeds, were reported to suppress papaya dieback disease caused by *Erwinia mallotivora*. The combined application of *Weissella cibaria* and *Lactococcus lactis* increased antibacterial activity against the pathogen and reduced disease severity. In citrus fruit, Azevedo et al. (2016) reported biocontrol activity of citrus endophyte *Methylobacterium mesophilicum* against *Xylella fastidiosa*, the causal agent of citrus variegated chlorosis (CVC) in sweet orange (*Citrus sinensis* L), and suggested that the endophytic microbial community of orange has a role in the pathogenicity of *Xylella fastidiosa*. Volatile organic compounds secreted from the endophytic fungal strains *of Nodulisporium* sp. and *Hypoxylon anthochroum* were reported to inhibit the post-harvest control of *Fusarium oxysporum* in cherry tomatoes. The application of fungal volatiles individually or in combination alters the cell membrane permeability as well as damage Fusarium hyphal morphology (Medina-Romero et al. 2017). Endophytic strain of *Pseudomonas synxantha* isolated from the tree of kiwi fruit (Aiello et al. 2019) exhibited antagonistic activity after harvest against *Monilinia fructicola* and *Monilinia fructigena*, the causal agents of brown rot of stone fruit. Significant growth inhibition of both the pathogens *in vitro* and on fruit was demonstrated.

Future Prospective

Microbial endophytes of the carposphere have been reported as an important component of the microbiome. These endophytes have potential as effective biocontrol agents as well as other uses for the regulation of fruit senescence and protection against physiological disorders. Endophytic microorganisms have an excellent capability to colonize different parts of fruit tissue and interact in multiple ways with each other and with the host. However, at this stage, there is a need to better understand the biology of fruit endophytes and elucidate their function and interaction with their host. Fruit endophytes would be a great source of bioactive compounds that have various biological functions.

References

Abdel-Rahim IR, Abo-Elyousr KA (2017) Using of endophytic Saccharomycopsis fibuligera and thyme oil for management of gray mold rot of guava fruits. Biol Control 110:124–131. https://doi.org/10.1016/j.biocontrol.2017.04.014

Aiello D, Restuccia C, Stefani E, Vitale A, Cirvilleri G (2019) Postharvest biocontrol ability of *Pseudomonas synxantha* against *Monilinia fructicola* and *Monilinia fructigena* on stone fruit. Postharvest Biol Tech 149:83–89. https://doi.org/10.1016/j.postharvbio.2018.11.020

Aleklett K, Hart M, Shade A (2014) The microbial ecology of flowers: an emerging frontier in phyllosphere research 1. Botany 92:253–266. https://doi.org/10.1139/cjb-2013-0166

Alkan N, Fortes AM (2015) Insights into molecular and metabolic events associated with fruit response to post-harvest fungal pathogens. Front Plant Sci 6:889. https://doi.org/10.3389/fpls.2015.00889

Al-Kharousi ZS, Guizani N, Al-Sadi AM, Al-Bulushi IM, Shaharoona B (2016) Hiding in fresh fruits and vegetables: opportunistic pathogens may cross geographical barriers. Int J Microbiol 2016:1–14. https://doi.org/10.1155/2016/4292417

Azevedo JL, Araújo WL, Lacava PT (2016) The diversity of citrus endophytic bacteria and their interactions with *Xylella fastidiosa* and host plants. Genet Mol Biol 39:476–491. https://doi.org/10.1590/1678-4685-gmb-2016-0056

Böhm M, Hurek T, Reinhold-Hurek B (2007) Twitching motility is essential for endophytic rice colonization by the N2-fixing endophyte *Azoarcus* sp. strain BH72. Mol Plant-Microbe Interact 20:526–533. https://doi.org/10.1094/MPMI-20-5-0526

Brader G, Compant S, Mitter B, Trognitz F, Sessitsch A (2014) Metabolic potential of endophytic bacteria. Curr Opin Biotechnol 27:30–37. https://doi.org/10.1016/j.copbio.2013.09.012

Bright M, Bulgheresi S (2010) A complex journey: transmission of microbial symbionts. Nat Rev Microbiol 8:218–230. https://doi.org/10.1038/nrmicro2262

Chanway CP (1996) Endophytes: they're not just fungi! Can J Bot 74:321–322

Compant S, Duffy B, Nowak J, Clément C, Barka EA (2005) Use of plant growth-promoting bacteria for biocontrol of plant diseases: principles, mechanisms of action, and future prospects. Appl Environ Microbiol 71:4951–4959. https://doi.org/10.1128/AEM.71.9.4951-4959.2005

Compant S, Kaplan H, Sessitsch A, Nowak J, Ait Barka E, Clément C (2008) Endophytic colonization of *Vitis vinifera* L. by *Burkholderia phytofirmans* strain PsJN: from the rhizosphere to inflorescence tissues. FEMS Microbiol Ecol 63:84–93. https://doi.org/10.1111/j.1574-6941.2007.00410.x

Compant S, Clément C, Sessitsch A (2010) Plant growth-promoting bacteria in the rhizo-and endosphere of plants: their role, colonization, mechanisms involved and prospects for utilization. Soil Biol Biochem 42:669–678. https://doi.org/10.1016/j.soilbio.2009.11.024

Compant S, Mitter B, Colli-Mull JG, Gangl H, Sessitsch A (2011) Endophytes of grapevine flowers, berries, and seeds: identification of cultivable bacteria, comparison with other plant parts, and visualization of niches of colonization. Microb Ecol 62:188–197. https://doi.org/10.1007/s00248-011-9883-y

Conway WS, Janisiewicz WJ, Leverentz B, Saftner RA, Camp MJ (2007) Control of blue mold of apple by combining controlled atmosphere, antagonist mixture, and sodium bicarbonate. Postharvest Biol Technol 45:326–332. https://doi.org/10.1016/j.postharvbio.2007.03.005

Cruz AF, Barka GD, Blum LEB, Tanaka T, Ono N, Kanaya S, Reineke A (2019) Evaluation of microbial communities in peels of Brazilian tropical fruits by amplicon sequence analysis. Braz J Microbiol 50:739–748. https://doi.org/10.1007/s42770-019-00187-y

De Bary A (1866) Morphologie und Physiologie der Pilze, Flechten, und Myxomyceten. Hofmeister's handbook of physiological Botany. W. Engelmann, Leipzig

De Melo Pereira GV, Magalhães KT, Lorenzetii ER, Souza TP, Schwan RF (2012) A multiphasic approach for the identification of endophytic bacterial in strawberry fruit and their potential for plant growth promotion. Microb Ecol 63:405–417. https://doi.org/10.1007/s00248-011-9919-3

Dong CJ, Wang LL, Li Q, Shang QM (2019) Bacterial communities in the rhizosphere, phyllosphere and endosphere of tomato plants. PLoS One 14:e0223847. https://doi.org/10.1371/journal.pone.0223847

Droby S, Wisniewski M (2018) The fruit microbiome: a new frontier for postharvest biocontrol and postharvest biology. Postharvest Biol Technol 140:107–112. https://doi.org/10.1016/j.postharvbio.2018.03.004

Droby S, Wisniewski M, Teixidó N, Spadaro D, Jijakli MH (2019) Biocontrol of postharvest diseases with antagonistic microorganisms. In: Postharvest pathology of fresh horticultural produce. CRC Press

Elbeltagy A, Nishioka K, Suzuki H, Sato T, Sato YI, Morisaki H, Mitsui H, Minamisawa K (2000) Isolation and characterization of endophytic bacteria from wild and traditionally cultivated rice varieties. Soil Sci Plant Nutr 46:617–629. https://doi.org/10.1080/00380768.2000.10409127

Elsen S, Swem LR, Swem DL, Bauer CE (2004) RegB/RegA, a highly conserved redox-responding global two-component regulatory system. Microbiol Mol Biol Rev 68:263–279. https://doi.org/10.1128/MMBR.68.2.263-279.2004

Frank AC, Saldierna Guzmán JP, Shay JE (2017) Transmission of bacterial endophytes. Microorganisms 5(4):70. https://doi.org/10.3390/microorganisms5040070

Galsurker O, Diskin S, Maurer D, Feygenberg O, Alkan N (2018) Fruit stem-end rot. Horticulturae 4:50. https://doi.org/10.3390/horticulturae4040050

Gasser I, Cardinale M, Müller H, Heller S, Eberl L, Lindenkamp N, Kaddor C, Steinbüchel A, Berg G (2011) Analysis of the endophytic lifestyle and plant growth promotion of *Burkholderia terricola* ZR2-12. Plant Soil 347:125. https://doi.org/10.1007/s11104-011-0833-8

Glassner H, Zchori-Fein E, Compant S, Sessitsch A, Katzir N, Portnoy V, Yaron S (2015) Characterization of endophytic bacteria from cucurbit fruits with potential benefits to agriculture in melons (*Cucumis melo L.*). FEMS Microbiol Ecol 91(7). https://doi.org/10.1093/femsec/fiv074

Hall ME, Wilcox WF (2019) Identification and frequencies of endophytic microbes within healthy grape berries. Am J Enol Vitic 70:212–219. https://doi.org/10.5344/ajev.2018.18033

Hallmann J, Quadt-Hallmann A, Mahaffee W, Kloepper J (1997) Bacterial endophytes in agricultural crops. Can J Microbiol 43:895–914. https://doi.org/10.1139/m97-131

Hanin NA, Fitriasari PD (2019) Identification of endophytic fungi from fruits and seeds of Jambolana (*Syzygium cumini* L.) Skeels. In: IOP conference series: Earth and Environmental Science (276: 012060). IOP Publishing

Hardoim PR, Van Overbeek LS, Berg G, Pirttilä AM, Compant S, Campisano A, Döring M, Sessitsch A (2015) The hidden world within plants: ecological and evolutionary considerations for defining functioning of microbial endophytes. Microbiol Mol Biol Rev 79:293–320. https://doi.org/10.1128/MMBR.00050-14

Hipol RM, Magtoto LM, Tamang SMA, Damatac AM II (2014) Antioxidant activities of fungal endophytes isolated from strawberry Fragaria× ananassa fruit. Electron J Biol 10:107–112

Istifadah N, Ningtyas DNY, Suryatmana P, Fitriatin BN (2017) The abilities of endophytic and biofertilizing bacteria and their combinations to suppress bacterial wilt disease (Ralstonia solanacearum) of chili. In: 2nd international conference on Sustainable Agriculture and Food Security: A Comprehensive Approach, KnE Life Sciences, pp 296–304

Iwai H, Usui M, Hoshino H, Kamada H, Matsunaga T, Kakegawa K, Ishii T, Satoh S (2003) Analysis of sugars in squash xylem sap. Plant Cell Physiol 44:582–587. https://doi.org/10.1093/pcp/pcg075

James EK, Gyaneshwar P, Mathan N, Barraquio WL, Reddy PM, Iannetta PP, Olivares FL, Ladha JK (2002) Infection and colonization of rice seedlings by the plant growth-promoting bacterium *Herbaspirillum seropedicae* Z67. MPMI 15:894–906. https://doi.org/10.1094/MPMI.2002.15.9.894

Janisiewicz WJ, Jurick WM II, Vico I, Peter KA, Buyer JS (2013) Culturable bacteria from plum fruit surfaces and their potential for controlling brown rot after harvest. Postharvest Biol Technol 76:145–151. https://doi.org/10.1016/j.postharvbio.2012.10.004

Johnson GI, Mead AJ, Cooke AW, Dean JR (1992) Mango stem end rot pathogens-Fruit infection by endophytic colonisation of the inflorescence and pedicel. Ann Appl Biol 120:225–234. https://doi.org/10.1111/j.1744-7348.1992.tb03420.x

Keller L, Surette MG (2006) Communication in bacteria: an ecological and evolutionary perspective. Nat Rev Microbiol 4:249–258. https://doi.org/10.1038/nrmicro1383

Krishnan P, Bhat R, Kush A, Ravikumar P (2012) Isolation and functional characterization of bacterial endophytes from Carica papaya fruits. J Appl Microbiol 113:308–317. https://doi.org/10.1111/j.1365-2672.2012.05340.x

Kumar A, Droby S, Singh VK, Singh SK, White JF (2020) Entry, colonization, and distribution of endophytic microorganisms in plants. In: Microbial Endophytes. Woodhead Publishing, pp 1–33. https://doi.org/10.1016/B978-0-12-819654-0.00001-6

Lai K, Chen S, Hu M, Hu Q, Geng P, Weng Q, Jia J (2012) Control of postharvest green mold of citrus fruit by application of endophytic *Paenibacillus polymyxa* strain SG-6. Postharvest Biol Technol 69:40–48. https://doi.org/10.1016/j.postharvbio.2012.03.001

Liu J, Abdelfattah A, Norelli J, Burchard E, Schena L, Droby S, Wisniewski M (2018) Apple endophytic microbiota of different rootstock/scion combinations suggests a genotype-specific influence. Microbiome 6(1):18. https://doi.org/10.1186/s40168-018-0403-x

Lodewyckx C, Vangronsveld J, Porteous F, Moore ER, Taghavi S, Mezgeay M, der Lelie DV (2002) Endophytic bacteria and their potential applications. Crit Rev Plant Sci 21:583–606. https://doi.org/10.1080/0735-260291044377

Lukša J, Vepštaitė-Monstavičė I, Apšegaitė V, Blažytė-Čereškienė L, Stanevičienė R, Strazdaitė-Žielienė Ž, Ravoitytė B, Aleknavičius D, Būda V, Mozūraitis R, Servienė E (2020) Fungal microbiota of sea buckthorn berries and its 2 volatile profiles at two ripening stages. Microorganisms 8:456. https://doi.org/10.3390/microorganisms8030456

Madbouly AK, Elyousr KAA, Ismail IM (2020) Biocontrol of Monilinia fructigena, causal agent of brown rot of apple fruit, by using endophytic yeasts. Biol Control 144:104239. https://doi.org/10.1016/j.biocontrol.2020.104239

Malfanova N, Lugtenberg BJ, Berg G (2013) Bacterial endophytes: who and where, and what are they doing there. Mol Microb Ecol Rhizosphere 1:15–37. Frans J de Bruijn

Medina-Romero YM, Roque-Flores G, Macías-Rubalcava ML (2017) Volatile organic compounds from endophytic fungi as innovative postharvest control of *Fusarium oxysporum* in cherry tomato fruits. Appl Microbiol Biotechnol 101:8209–8222. https://doi.org/10.1007/s00253-017-8542-8

Miguel PSB, Delvaux JC, Oliveira MNV, Monteiro LCP, Freitas FDS, Costa MD, Tótola MR, Moraes CA, Borges AC (2013) Diversity of endophytic bacteria in the fruits of Coffea canephora. Afr J Microbiol Res 7:586–594. https://doi.org/10.5897/AJMR12.2036

Nigris S, Baldan E, Tondello A, Zanella F, Vitulo N, Favaro G, Guidolin V, Bordin N, Telatin A, Barizza E, Marcato S (2018) Biocontrol traits of Bacillus licheniformis GL174, a culturable endophyte of Vitis vinifera cv. Glera. BMC Microbiol 18(1):133. https://doi.org/10.1186/s12866-018-1306-5

Oliveira MN, Santos TM, Vale HM, Delvaux JC, Cordero AP, Ferreira AB, Miguel PS, Tótola MR, Costa MD, Moraes CA, Borges AC (2013) Endophytic microbial diversity in coffee cherries of *Coffea arabica* from southeastern Brazil. Can J Microbiol 59:221–230. https://doi.org/10.1139/cjm-2012-0674

Ottesen AR, Peña AG, White JR, Pettengill JB, Li C, Allard S, Rideout S, Allard M, Hill T, Evans P, Strain E (2013) Baseline survey of the anatomical microbial ecology of an important food plant: Solanum lycopersicum (tomato). BMC Microbiol 13:114. https://doi.org/10.1186/1471-2180-13-114

Ou T, Xu WF, Wang F, Strobel G, Zhou ZY, Xiang ZH, Liu J, Xie J (2019) A microbiome study reveals seasonal variation in endophytic bacteria among different mulberry cultivars. CSBJ 17:1091–1100. https://doi.org/10.1016/j.csbj.2019.07.018

Petrini O (1991) Fungal endophytes of tree leaves. In: Microbial ecology of leaves. Springer, New York, pp 179–197

Prakamhang J, Minamisawa K, Teamtaisong K, Boonkerd N, Teaumroong N (2009) The communities of endophytic diazotrophic bacteria in cultivated rice (*Oryza sativa L.*). Appl Soil Ecol 42:141–149. https://doi.org/10.1016/j.apsoil.2009.02.008

Prusky D, Alkan N, Mengiste T, Fluhr R (2013) Quiescent and necrotrophic lifestyle choice during postharvest disease development. Annu Rev Phytopathol 51:155–176. https://doi.org/10.1146/annurev-phyto-082712-102349

Rasche F, Lueders T, Schloter M, Schaefer S, Buegger F, Gattinger A, Hood-Nowotny RC, Sessitsch A (2009) DNA-based stable isotope probing enables the identification of active bacterial endophytes in potatoes. New Phytol 181:802–807. https://doi.org/10.1111/j.1469-8137.2008.02744.x

Reinhold-Hurek B, Maes T, Gemmer S, Van Montagu M, Hurek T (2006) An endoglucanase is involved in infection of rice roots by the not-cellulose-metabolizing endophyte *Azoarcus* sp. strain BH72. Mol Plant-Microbe Interact 19:181–188. https://doi.org/10.1094/MPMI-19-0181

Ren F, Dong W, Yan DH (2019a) Endophytic bacterial communities of Jingbai Pear trees in north China analyzed with Illumina sequencing of 16S rDNA. Arch Microbiol 201:199–208. https://doi.org/10.1007/s00203-018-1597-9

Ren F, Dong W, Sun H, Yan DH (2019b) Endophytic Mycobiota of Jingbai pear trees in north China. Forests 10:260. https://doi.org/10.3390/f10030260

Rosenblueth M, Martinez-Romero E (2006) Bacterial endophytes and their interaction with hosts. Mol Plant-Microbe Interact 19:827–837. https://doi.org/10.1094/MPMI-19-0827

Saikkonen K, Wäli P, Helander M, Faeth SH (2004) Evolution of endophyte–plant symbioses. Trends Plant Sci 9:275–280. https://doi.org/10.1016/j.tplants.2004.04.005

Saminathan T, García M, Ghimire B, Lopez C, Bodunrin A, Nimmakayala P, Abburi VL, Levi A, Balagurusamy N, Reddy UK (2018) Metagenomic and metatranscriptomic analyses of diverse watermelon cultivars reveal the role of fruit associated microbiome in carbohydrate metabolism and ripening of mature fruits. Front Plant Sci 9:4. https://doi.org/10.3389/fpls.2018.00004

Sauer K, Camper AK (2001) Characterization of phenotypic changes in *Pseudomonas putida* in response to surface-associated growth. J Bacteriol 183:6579–6589. https://doi.org/10.1128/JB.183.22.6579-6589.2001

Schena L, Nigro F, Pentimone I, Ligorio A, Ippolito A (2003) Control of postharvest rots of sweet cherries and table grapes with endophytic isolates of *Aureobasidium pullulans*. Postharvest Biol Technol 30:209–220. https://doi.org/10.1016/S0925-5214(03)00111-X

Shameer S, Prasad TNVKV (2018) Plant growth promoting rhizobacteria for sustainable agricultural practices with special reference to biotic and abiotic stresses. Plant Growth Regul 84:603–615. https://doi.org/10.1007/s10725-017-0365-1

Shi J, Liu A, Li X, Feng S, Chen W (2010) Identification of endophytic bacterial strain MGP1 selected from papaya and its biocontrol effects on pathogens infecting harvested papaya fruit. J Sci Food Agric 90(2):227–32. https://doi.org/10.1002/jsfa.3798

Shimizu M, Yazawa S, Ushijima Y (2009) A promising strain of endophytic Streptomyces sp. for biological control of cucumber anthracnose. J Gen Plant Pathol 75:27–36. https://doi.org/10.1007/s10327-008-0138-9

Singh VK, Singh AK, Kumar A (2017) Disease management of tomato through PGPB: current trends and future perspective. Biotech 7:255. https://doi.org/10.1007/s13205-017-0896-1

Singh M, Singh D, Gupta A, Pandey KD, Singh PK, Kumar A (2019) Plant growth promoting rhizobacteria: application in biofertilizers and biocontrol of phytopathogens. In: PGPR amelioration in sustainable agriculture, vol 3. Woodhead Publishing, pp 41–66. https://doi.org/10.1016/B978-0-12-815879-1.00003-3

Singh M, Srivastava M, Kumar A, Singh AK, Pandey KD (2020) Endophytic bacteria in plant disease management. In: Microbial endophytes, vol 4. Woodhead Publishing, pp 61–89. https://doi.org/10.1016/B978-0-12-818734-0.00004-8

Sritharan T, Savitri Kumar N, Jayasinghe L, Araya H, Fujimoto Y (2019) Isocoumarins and dihydroisocoumarins from the endophytic fungus biscogniauxia capnodes isolated from the fruits of averrhoa carambola. Nat Prod Commun 14:1934578X19851969. https://doi.org/10.1177/1934578X19851969

Sturz AV, Christie BR (1996) Endophytic bacteria of red clover as causal agents of allelopathic clover–maize syndromes. Soil Biol Biochem 28:583–588. https://doi.org/10.1016/0038-0717(95)00168-9

Suárez-Moreno ZR, Devescovi G, Myers M, Hallack L, Mendonça-Previato L, Caballero-Mellado J, Venturi V (2010) Commonalities and differences in regulation of N-acyl homoserine lactone quorum sensing in the beneficial plant-associated *Burkholderia* species cluster. Appl Environ Microbiol 76:4302–4317. https://doi.org/10.1128/AEM.03086-09

Suhandono S, Kusumawardhani MK, Aditiawati P (2016) Isolation and molecular identification of endophytic bacteria from Rambutan fruits (*Nephelium lappaceum L.*) cultivar Binjai. HAYATI J Biosci 23:39–44. https://doi.org/10.1016/j.hjb.2016.01.005

Sy A, Timmers ACJ, Knief C, Vorholt JA (2005) Methylotrophic metabolism is advantageous for *Methylobacterium extorquens* during colonization of Medicago truncatula under competitive conditions. Appl Environ Microbiol 71:7245–7252. https://doi.org/10.1128/AEM.71.11.7245-7252.2005

Wu W, Lei J, Hussain M, Cao S, Du B, Wang R (2019) Structure and function of the fruit microbiome in healthy and diseased kiwifruit. Pakistan J Agric Sci 56:3. https://doi.org/10.21162/PAKJAS/19.8820

Zheng H, Mao Y, Teng J, Zhu Q, Ling J, Zhong Z (2015) Flagellar-dependent motility in *Mesorhizobium tianshanense* is involved in the early stage of plant host interaction: study of an flgE mutant. Curr Microbiol 70:219–227. https://doi.org/10.1007/s00284-014-0701-x

Zinniel DK, Lambrecht P, Harris NB, Feng Z, Kuczmarski D, Higley P, Ishimaru CA, Arunakumari A, Barletta RG, Vidaver AK (2002) Isolation and characterization of endophytic colonizing bacteria from agronomic crops and prairie plants. Appl Environ Microbiol 68:2198–2208. https://doi.org/10.1128/AEM.68.5.2198-2208.2002

Chapter 6
The Production of Mycotoxins as an Adaptation to the Post-Harvest Environment

Rolf Geisen and Markus Schmidt-Heydt

Abstract Many saprophytic and plant pathogenic environmental fungi can produce secondary metabolites which consist of a plethora of different chemicals with very specific structures. Secondary metabolites are usually not necessary for growth especially on laboratory media. Among secondary metabolites, mycotoxins are toxic against humans and animals. From a biological point of view this toxicity of course is not the ecological reason for the biosynthesis of these toxins by fungi, because, except for the human and animal pathogenic ones, there is in general no direct ecological niche overlap between humans or animals and saprophytic and plant pathogenic filamentous fungi. The biological role for the synthesis of mycotoxins is still not completely clear. However recently more and more insights into this biological question become obvious. Mycotoxins are described as pathogenicity factors of plants, colonization factors in the post-harvest environment or as specific adaptation factors for certain more challenging and stressful environments. Especially environmental stresses lead to induction of mycotoxin biosynthesis. Most of these stress inductions are guided by signal transduction pathways which regulate the transcription of the mycotoxin biosynthesis genes in relation to environmental changes. In this chapter an overview about the interplay of the post-harvest environment and mycotoxin biosynthesis will be given.

Keywords Adaptation · Mycotoxin · Aflatoxin · Ochratoxin · Patulin · Alternariol · Molecular regulation · Stress · Competitiveness · Colonization · Post-harvest

---

R. Geisen (✉) · M. Schmidt-Heydt
Department of Safety and Quality of Fruit and Vegetables, Max Rubner-Institut, Karlsruhe, Germany
e-mail: rolf.geisen@mri.bund.de; markus.schmidt-heydt@mri.bund.de

D. Spadaro et al. (eds.), *Postharvest Pathology*, Plant Pathology in the 21st Century 11, https://doi.org/10.1007/978-3-030-56530-5_6

Introduction

It is estimated that about 25% of the annual harvest of plants, including fruits and vegetables, are spoiled by fungi (Bhatnagar et al. 2002; Gomez et al. 2015). Many of these fungi can produce mycotoxins, which further aggravate the situation. Historically, in mycotoxin research, there is a differentiation between field fungi (pathogenic fungi) and storage fungi (saprophytic fungi). Field fungi have real pathogenic activities and can actively infect plants on the field, e. g. during plant growth when plant defense systems can be fully activated. Storage fungi usually occur and grow in the post-harvest environment. Under these conditions the protective functions of the harvested plant parts are reduced. Nevertheless, harvested fruits and vegetables are still able to defend against invading pathogens. For example, salicylic acid and jasmones, endogenous metabolites and regulators can have an influence on the resistance of the plant against fungal attack. According to Asghari and Aghdam (2010) such metabolites can activate defense genes by influencing the ROS (reactive oxygen species) concentration of the cell. Generally, an increase in the ROS concentration constitutes a protection against invading fungi. (Mittler 2012). Liu et al. (2018) pointed out that the viability and virulence of decay fungi is dependent on the resilience of the fungi against a set of different environmental conditions like heat stress, pH or oxidative stress. These parameters do not only play a role in the viability and the virulence of the decay fungi, but also on mycotoxin biosynthesis by toxigenic fungi (Magan et al. 2003). Similarly, Alkan and Fortes (2015) reviewed the interplay between the post-harvest environment, ripening, host response and fungal pathogenicity. They discussed that certain host factors like the jasmonate – salicylate crosstalk or the activity of ethylene also play important roles whether a fungus can establish a colonization on a fruit or vegetable in the post-harvest state. There is not always a clear differentiation between a field fungus (pathogenic state) and a storage fungus (saprophytic state). *Aspergillus flavus* for example occurs and produces aflatoxins both during growth on maize in the field and during post-harvest (Sanzani et al. 2016). During growth in the field, where the water activity may be high, other fungi like *Fusarium* spp. may have growth advantages, however during the post-harvest stage when the grains dry, *Aspergillus* spp. may overgrow other fungi because of their tolerance to lower content of free water and may lead to ample concentrations of aflatoxins. That means that silent field infections may switch to high aflatoxin producing infections during the storage phase of the fungus. This behavior suggests some pathogenic state of *A. flavus*, when infecting maize plants on the field and a more saprophytic state during growth under storage conditions. According to Sanzani et al. (2016) especially the latter is accompanied by high production of aflatoxin.

Some important filamentous fungi of the post-harvest environment which produce mycotoxins are listed in Table 6.1. These include members of the genera *Aspergillus*, *Penicillium*, and *Alternaria*. Species of the genus *Aspergillus* are mainly responsible for the production of aflatoxins, cyclopiazonic acid and ochratoxin. Species of the genus *Penicillium* may produce ochratoxin, citrinin or patulin,

Table 6.1 Important mycotoxigenic fungi of the post-harvest environment

Species	Mycotoxins	Habitat	Parameter
Aspergillus flavus	Aflatoxin, cyclopiazonic acid	Maize, nuts, figs	Oxidative stress, temperature
Aspergillus ochraceus	Ochratoxin	Grapes	Temperature, pH
Penicillium expansum	Patulin, citrinin	Appels, other fruits	Oxidative stress, pH
Penicillium verrucosum	Ochratoxin, citrinin	Wheat, salt fermented olives	Light/dark conditions, oxidative stress, NaCl stress
Alternaria alternata (saprophytic strains)	Alternariol, tenuazonic acid	Tomatoes, fruits, vegetables	Water activity, osmotic stress, pH

whereas species of the genus *Alternaria* are able to produce alternariol, alternariol monomethylether or tenuazonic acid as the most important post-harvest mycotoxins. Less frequent mycotoxin producers in the post-harvest environment are species of the genus *Byssochlamys* or *Paecilomyces* able to produce patulin. The important spoilage fungi of the genus *Botrytis* and representatives of the *Mucoraceae* as well as further representatives of the genus *Penicillium*, like *P. digitatum* and *P. italicum*, are not regarded as typical mycotoxin producing fungi.

There is no clear-cut explanation why fungi produce mycotoxins. It is well known that mutant strains, which are not able to produce the respective mycotoxins, grow equally well as the wild type, at least on laboratory media. In hypotheses found in the recent literature, more and more the view that the production of mycotoxins (and other secondary metabolites) support the adaptation of the fungus to a certain environment, especially under certain stress conditions, is being described (Reverberi et al. 2010). In the post-harvest environment, the fungus must adapt to the specific conditions, which are not always favorable for the fungus. The production of mycotoxins may be regarded as one factor to support this niche adaptation (Chalivendra et al. 2017).

Influence of Stress Conditions on Mycotoxin Biosynthesis

During post-harvest storage of fruits and vegetables the conditions regarding humidity/water activity and temperature should be controlled in a way that fungal spoilage is avoided or at least reduced due to the low temperature or the reduced humidity/water activity. The safest situation in this respect would be a complete inhibition of fungal growth. However, achieving this goal requires high energy consumption especially because fungi can grow at low temperatures and at low water activity values. This leads to the situation that, under post-harvest storage conditions, slow marginal growth of the fungus may still be possible. Growth under these conditions means stress for the fungus (Schmidt-Heydt et al. 2015). It has long been noted that various stress conditions lead to an induction of mycotoxin biosynthesis. This is also true for temperature and water activity stresses. Schmidt-Heydt et al. (2008)

showed that expression of ochratoxin biosynthesis genes and thereby ochratoxin biosynthesis in *P. verrucosum* is activated in two peaks depending on temperature, water activity or pH. One high peak of ochratoxin biosynthesis could be identified around the optimum conditions for growth and one lower peak appeared close to the margins of growth. This was demonstrated by continuously changing two parameters (temperature, water activity) in a bifactorial experiment. The expression of the ochratoxin related biosynthesis genes were monitored by microarray (Fig. 6.1). A high peak of expression of the ochratoxin biosynthesis genes and thereby ochratoxin biosynthesis could be identified at optimal growth conditions and a second minor peak could be found near the margin of growth (Schmidt-Heydt et al. 2008). This microarray experiment very clearly shows the two different optima for mycotoxin biosynthesis. This was also true for various other mycotoxin producing species like for aflatoxin biosynthesis by *A. parasiticus* and trichothecene biosynthesis by *F. culmorum* (Schmidt-Heydt et al. 2008). Similarly, Gallo et al. (2016) found that moderate stress conditions, due to temperature and water activity, lead to increased production of aflatoxin by *A. flavus* on almond medium at a prolonged

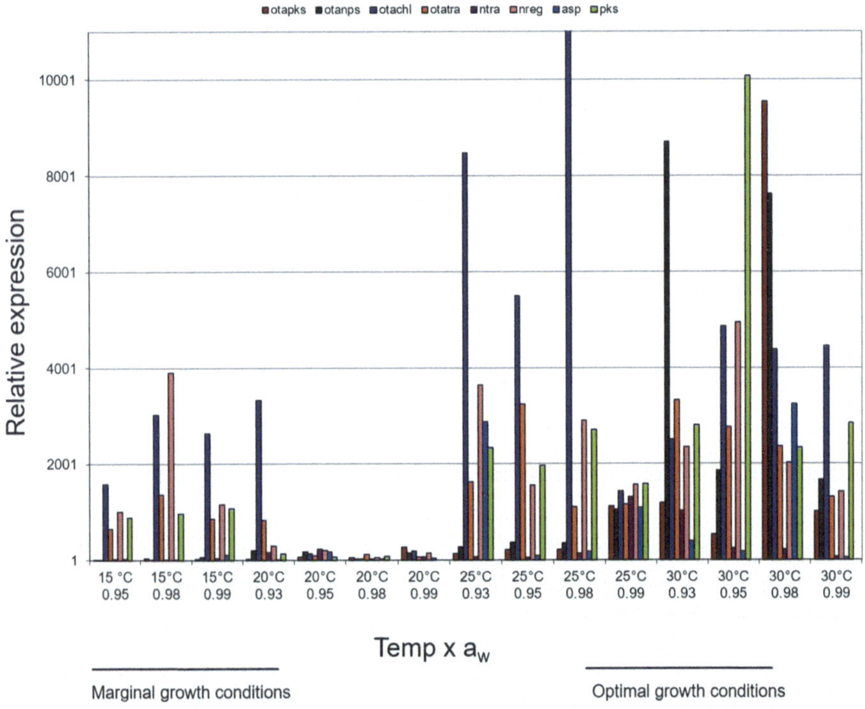

Fig. 6.1 Microarray expression data of ochratoxin A related genes of *P. verrucosum* grown under different combinations of temperature (Temp) and water activity (a_w). The marginal and optimal growth conditions are indicated, as well as the two expression peaks, which were paralleled by higher production of ochratoxin. The different bars represent the expression of genes co-regulated under ochratoxin biosynthesis conditions. Key genes are the *otapks* and *otanps* genes (ochratoxin polyketide synthase and ochratoxin non-ribosomal peptide synthase genes)

incubation time. Interestingly the regulatory genes *afl*R and *afl*S where more activated at these conditions compared to the structural genes *afl*D and *afl*O. Oviedo et al. (2010) analysed the influence of different combinations of water activity and temperature on the biosynthesis of alternariol by *A. alternata*. At a given temperature they also found different peaks of alternariol biosynthesis at water activities optimal and sub-optimal for growth. Because of this additional peak of mycotoxin biosynthesis under stress conditions, the production of mycotoxins under these conditions seems to have supporting and adaptive functions for growth. In any case the induction of mycotoxin biosynthesis under reduced temperature or water activity stress is an important issue in terms of food safety of stored agricultural products because they are often stored at exactly these conditions to prolong shelf life.

Oxidative stress due to an increased oxidative status in fruits and vegetables also can have a great influence on mycotoxin biosynthesis. Jayashree and Subramanyam (2000) for example showed that oxidative stress is a pre-requisite for the induction of aflatoxin biosynthesis by *A. parasiticus*. The relation between oxidants and antioxidants seems to play an important role for the regulation of aflatoxin biosynthesis by *A. parasiticus* (Reverberi et al. 2006) and other fungi and mycotoxins (Schmidt-Heydt et al. 2015). Reverberi et al. (2012) showed that also the production of ochratoxin by *A. ochraceus* was induced after growth on oxidant containing medium. Moreover, the deletion of the *Aoyap*1 gene, a transcription factor regulated by the oxidative status of the habitat, leads to changes in ochratoxin biosynthesis. In fact, in a mutant strain a hyperoxidant status was mimicked which leads to the production of higher amounts of ochratoxin compared to the wild type. Upon fungal attack plants react and increases their internal ROS status also in the post-harvest phase (Aver'yanov et al. 2012). On the other side also fungi can regulate their internal ROS concentration to support the colonization of the plant matrix. Therefore, the colonizing fungus as well as the plant host balances the ROS concentration versus the antioxidative activity in a way that the own cells are protected as far as possible but that either defense or attack is still functional. Counterparts of the reactive oxygen species are some enzymatic systems like superoxide dismutases, catalases, glutathione- or ascorbic dependent peroxidases or protective systems like melanines, carotenes, ascorbate, proline or mechanisms like chelation of ferrous ions by siderophores or scavenging of hydroxyl or other free radicals (Aver'yanov et al. 2012). However, also mycotoxins may have protective or antioxidative activities which can play a role in the post-harvest plant environment and defend against oxidative attack (Heider et al. 2006). In contrast, fungal toxins can also have impacts on the increase of ROS in the host plants. Tenuazonic acid for example, a mycotoxin of *A. alternata*, induces ROS production in treated rice plants, thereby increasing the resistance to the fungal attack (Aver'yanov et al. 2007). Similarly, the trichothecene desoxynivalenol, produced by *Fusarium* spp., increases the level of H_2O_2 in wheat leaves (Desmond et al. 2008). These facts show that there is a dedicated cross talk between the host plant and the attacking fungus, also in the post-harvest situation, and that secondary metabolites like mycotoxins play an important role in that communication.

Aflatoxin and Cyclopiazonic Acid Production by *Aspergillus flavus*

Aspergillus flavus is, beside *A. parasiticus*, one of the most important aflatoxin producing fungus. Aflatoxin can be found in Brazil nuts (Massi et al. 2014), peanuts (Palencia et al. 2010), maize (Reverberi et al. 2013) and figs (Iamanaka et al. 2007) among other commodities. The molecular background of aflatoxin biosynthesis is well known. The biosynthesis genes are clustered on an 80 kb long genomic region on chromosome 3 (Yu et al. 2004; Chang et al. 2009a). The cluster for cyclopiazonic acid is located just in the vicinity of the aflatoxin gene cluster (Chang et al. 2009a; Chang and Ehrlich 2011) (Fig. 6.2). The intergenic region between the aflatoxin and the cyclopiazonic acid gene cluster contains some regulatory elements. Many *A. flavus* strains possess deletions in that region (Chang et al. 2009b). Because of frequent deletions in that region *A. flavus* strains isolated from nature have varying capacities to produce either aflatoxin or cyclopiazonic acid or both. Nevertheless, these secondary metabolites may play a role during growth in the field or during post-harvest conditions. As mentioned above *A. flavus* is a weak plant pathogenic but mainly a saprophytic fungus (Zhao et al. 2018). Reverberi et al. (2013) analysed the shift from pathogenic to saprophytic growth. By a transcriptomics approach they observed a change in gene expression, especially for secondary metabolite gene clusters. The authors found that the oxidative conditions are especially important during interaction with the host plant and during change of the life style, e. g. a change from saprophytic and pathogenic growth and *vice-versa*. During their transcriptomic analysis, the authors found that the genes in both the aflatoxin and cyclopiazonic acid clusters are co-expressed. The authors suggest a common regulatory mechanism for both clusters. Also, Chang et al. (2009b) suggests that both clusters may be co-regulated by *afl*R, the main regulatory gene of the aflatoxin biosynthesis cluster. According to various literature reports, *A. flavus* produces higher amounts of aflatoxin during the saprophytic post-harvest phase compared to the pathogenic field phase (Sanzani et al. 2016), which is an indication that these higher amounts may be important under these conditions. This view is supported by the work of Chalivendra et al. (2017). Theses authors described that aflatoxin biosynthesis under plant infection conditions is much more variable than during saprophytic growth.

Also plant products in the post-harvest phase increases their ROS concentrations after fungal attack (Touhami et al. 2018). This leads to a physiological reaction by

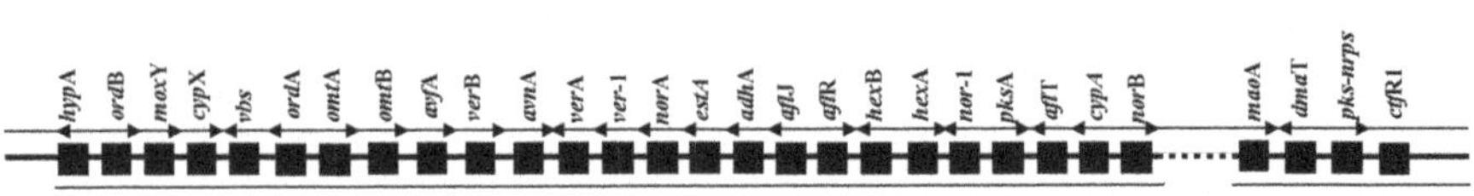

Fig. 6.2 Genetic organization of the aflatoxin (*afl*) and cyclopiazonic acid (*cpa*) gene clusters. The scheme is not drawn to scale(adapted according to Chang and Ehrlich (2011))

the fungus. According to Hong et al. (2013) it was proposed for *Aspergillus* that, during an antioxidative response, first antioxidative enzymes are activated and in a second responsive step secondary metabolites, like aflatoxins, are involved. This view is also supported in the review of Roze et al. (2013). In fact, external antioxidants, like ascorbic acid or plant inherent polyphenols usually results in the reduction of aflatoxin biosynthesis (Reverberi et al. 2005; Mahoney et al. 2010) suggesting that the presence of aflatoxin is no longer needed by the fungus.

Various strains of *A. flavus* can produce the mycotoxin cyclopiazonic acid in addition to aflatoxins. Compared to aflatoxins the toxicity of cyclopiazonic acid, with respect to humans and animals is much lower, however defined. Cyclopiazonic acid disturbs calcium uptake, which results in muscle disfunction (Chang et al. 2009b). Cyclopiazonic acid can also be produced by *Pencillium* spp., for example by *P. camemberti* (Le Bars 1979) or *P. griseofulvum* (Reddy and Reddy 1992). For that reasons it can be found in cheeses, especially when the cheese is stored at ambient temperatures (Le Bars 1979; Finoli et al. 1999; Hymery et al. 2014). Cyclopiazonic acid can also be found in peanuts (Lansden and Davison 1983) apparently produced by *A. flavus*. Also, Pinto et al. (2001) found aflatoxin and cyclopiazonic acid in peanuts grown in Argentina. In the work of Pinto et al. (2001) and in the work of Resnik et al. (1996) very high amounts of CPA were found in cereals. In the recent work of Chalivendra et al. (2017) it could be shown that cyclopiazonic acid apparently is a pathogenicity factor during the colonization of the living maize plant. As described above it disturbs calcium homeostasis and can thereby inactivate host defense response, which enables *A. flavus* to infect the plant. Mutants in cyclopiazonic acid biosynthesis were less virulent than the wild type (Chalivendra et al. 2017). The fact that cyclopiazonic acid has an influence on calcium homeostasis resembles the situation of ochratoxin production by *Penicillium*, which seems to have an influence on chloride homeostasis and is described below. Chang et al. (2009b) described a further biological reason for the biosynthesis of cyclopiazonic acid. Cyclopiazonic acid is a good chelator of iron and production of cyclopiazonic acid is, according to Chang et al. (2009b) usually high just before dormancy, e. g. just before the stop of growth. Fungi are often in an iron limited environment (for example soils) and it might be of advantage that they already have iron available when the conditions change, so they can immediately start to grow. Cyclopiazonic acid might therefore have a siderophore-like function (Chang et al. 2009b). According to Aver'yanov (2012), an accumulation of iron supports the antioxidative conditions in the cellular environment for example by enhancing catalase activity, which again would favor colonization of plants.

Whether the biosynthesis of aflatoxin and cyclopiazonic acid is somehow timely or environmentally correlated, which is the case for some of the examples below, is not yet systematically analysed.

Ochratoxin and Citrinin Production by *P. verrucosum*

P. verrucosum is one of the most important storage fungi of cereal grains (Cabanas et al. 2008; Dhungana et al. 2018). It has no known pathogenic activity on the plants in the field however contamination by spores of *P. verrucosum* obviously happens here (Elmholt and Hestbjerg 1999; Frisvad et al. 2005). At these conditions the fungal cells are partly exposed to bright sun light. After harvest and threshing the cereal grains are dried and stored in silos or store-houses, if possible at a controlled humidity in the dark. *P. verrucosum* is able to produce the two important mycotoxins ochratoxin and citrinin. Both are polyketides and are structurally related. *P. verrucosum* is the typical producing organism of ochratoxin and citrinin in cereals grown under moderate climatic conditions. Ochratoxigenic *Aspergillus* spp., like *A. niger*, *A. carbonarius* or *A. ochraceus* and the related species *A. steynii* and *A. westerdijkiae* usually do not play a role in this environment (Elmholt 2003; Frisvad et al. 2005). The presence of *P. verrucosum* in cereal grains at a certain frequency indicates the presence of ochratoxin. Lund and Frisvad (2003) showed that ochratoxin containing wheat samples had a much higher incidence of contamination by *P. verrucosum*, than cereal samples in which no ochratoxin could be detected. Lindblad et al. (2004) developed a mathematical model, which relates the level of *P. verrucosum* in grain and the respective water activity with the probability of the production of ochratoxin. A concentration of 1000 cfu/g was described to be an alert value indicating the production of ochratoxin. Interestingly grain isolated directly from the field can contain high amounts of citrinin (Pleadin et al. 2016). In fact, citrinin can often be found in higher quantities compared to ochratoxin (Vrabcheva et al. 2000; Wawrzyniak and Waśkiewicz 2014). In stored grain, according to the actual data, ochratoxin plays a more important role (Tittlemier et al. 2014) and might be higher than the concentrations of citrinin (Limay-Rios et al. 2017). That suggests that *P. verrucosum* might adapt mycotoxin biosynthesis depending on the outer environment producing more citrinin under field conditions and more ochratoxin under storage conditions in the dark. This view is supported by the fact that citrinin may have light protecting properties, which was already suggested by Størmer et al. (1998). Citrinin producing *Penicillium* species are more resistant against short wavelength light than citrinin non-producing species (Schmidt-Heydt et al. 2015). Citrinin is degraded by short wave blue light, which indicates that this molecule indeed absorbs energy from light, thereby protecting the producer (Schmidt-Heydt et al. 2012a). Moreover, it was shown that light also has an important influence on the biosynthesis of ochratoxin. Higher amounts of ochratoxin are produced by *P. verrucosum* after growth in the dark compared to growth under light conditions. It could be shown that higher amounts of ochratoxin apparently are toxic to the production organism itself, especially under light conditions. As a consequence, ochratoxin producing *Penicillium* spp. down-regulate their ochratoxin production under light conditions (Schmidt-Heydt et al. 2010). Irradiation by light of short wavelength can completely inhibit ochratoxin biosynthesis (Schmidt-Heydt et al. 2015), whereas the production of citrinin is induced under these conditions.

Light, being an electromagnetic radiation, can produce radicals and thereby increase the concentration of ROS in the cell (Trzaska et al. 2016; El-Esawi et al. 2017). Interestingly citrinin is not only regarded as a sun protectant, but also as an antioxidant (Ishikawa et al. 1988; Heider et al. 2006). In fact, also increased concentrations of oxidants in the medium (e. g. H_2O_2) leads to an increase in citrinin biosynthesis and a concomitant reduction of the ROS concentration in the treated fungal cells (Schmidt-Heydt et al. 2015; Touhami et al. 2018). The facts described above suggest that in the wheat environment the fungus regulates toxin biosynthesis according to the actual conditions. During contamination in the field, when the *P. verrucosum* spores are exposed to day light conditions, obviously more citrinin is produced, whereas under post-harvest conditions, when the grain is stored in the dark, the production of ochratoxin is preferred. Citrinin seems to be a protectant molecule under the conditions described. However, the biological reason why the fungus produces ochratoxin after growth on wheat in the dark is not yet clear (Fig. 6.3a). Perhaps it might favor the viability of the fungus in that environment. In fact, this is suggested by Maor et al. (2017) for the growth of *A. carbonarius* on grapes. They could show that conditions which reduce ochratoxin biosynthesis on grapes also reduces the colonization capacity.

There is another shift of both toxins in *P. verrucosum*, which is related to the specific adaptation to different post-harvest environments. As mentioned above the main habitat of *P. verrucosum* are cereals (Lund and Frisvad 2003). However, *P. verrucosum* can also be found on salt rich fermented vegetable foods like olives (Heperkan et al. 2009) and therefore, citrinin and ochratoxin can be found in these products. Tokuşoğlu and Bozoğlu (2010) found a high frequency of both toxins in processed olives. Seventy seven percent of the analysed black table olives contained citrinin and 94% of the samples contained ochratoxin. Also, El Adlouni et al. (2006) could identify ochratoxin in black greek olives at a high incidence. All of the analysed samples were positive for this toxin. Olives may be stored at non-controlled ambient temperatures (18–28 °C) for a longer time before processing. During that phase, beside the *Penicillium* spp. apparently *A. ochraceus* is described as a potential ochratoxin producing species (Ghitakou et al. 2006; Samane et al. 1991). The occurrence of high amounts of ochratoxin in olive oil may therefore be mainly due to the presence of both ochratoxigenic *Aspergillus* spp. and to a lesser extend to *P. verrucosum* (Papachristou and Markaki 2004; Ferracane et al. 2007). Processing of native olives to table olives however includes a salting step (Fernández et al. 1997). After de-bittering by NaOH the olives are fermented for up to several months in a brine solution, which, depending on the particular process may contain up to 10% NaCl (Hurtado et al. 2009; Abriouel et al. 2011). During this static fermentation moulds may grow on the surface of the brine solution. According to an analysis of Bavaro et al. (2017) they belong nearly exclusively to the genus *Penicillium*. *Aspergillus* is usually not that resistant to high concentrations of sodium chloride such as *A. carbonarius which is* not able to growth at 10% NaCl; however, *Penicillium species*, including *P. verrucosum* and *P.citrinum* are much better adapted to a high NaCl environment (Stoll et al. 2013). There is obviously a change in the ochratoxigenic fungal population from post-harvest olives (*A. ochraceus* and other

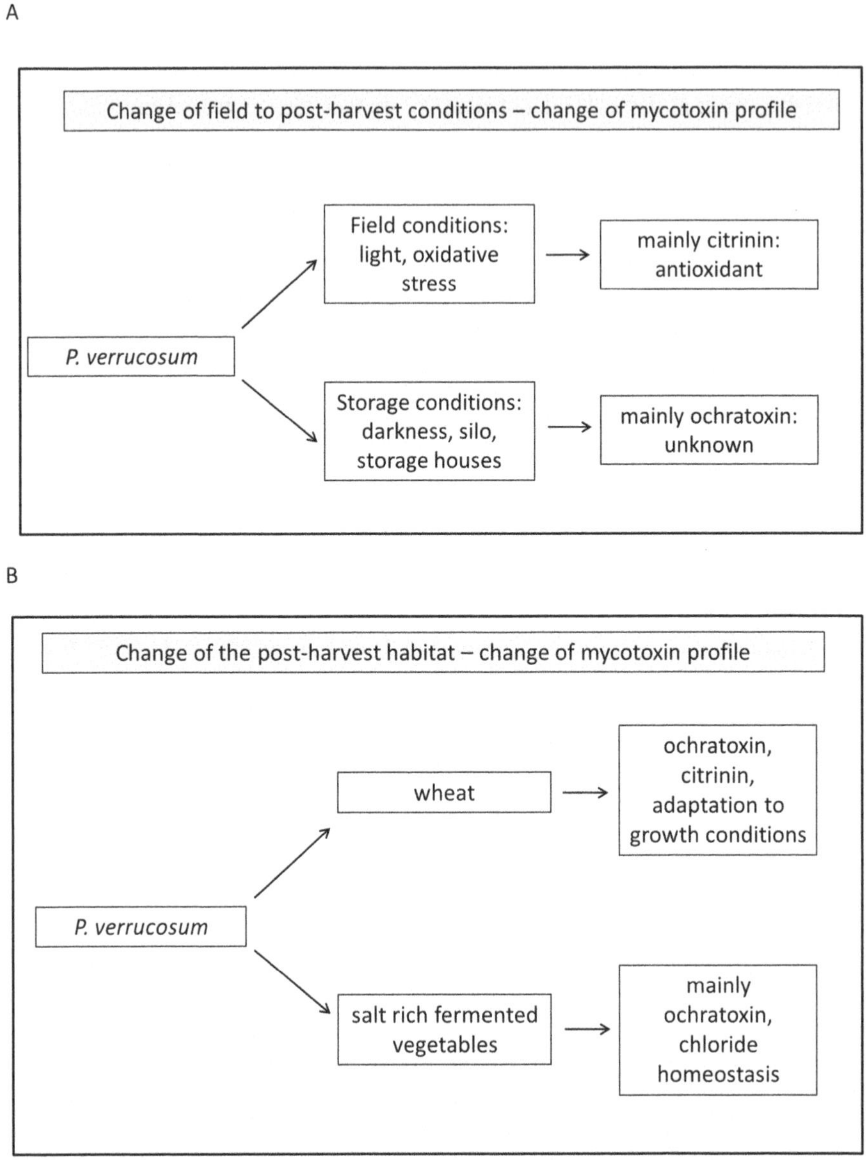

Fig. 6.3 Influence of growth conditions (**a**) or different habitats (**b**) on the mycotoxin production profile of *P. verrucosum*

Aspergillus spp.) to NaCl processed olives (*P. verrucosum* and *P. citrinum*). In fact, increasing concentrations of NaCl changes also the profile of secondary metabolites produced by *P. verrucosum*. Citrinin is usually produced by this species in advance to ochratoxin at zero to low concentrations of NaCl in the growth medium (potato dextrose agar), whereas increasing concentrations of NaCl shift the production towards ochratoxin. Schmidt-Heydt et al. (2012b) could show that the production of citrinin by *P. verrucosum* was completely inhibited and that only ochratoxin was produced at a NaCl concentration of 10%. In a molecular analysis of the high-osmolarity glycerol (HOG) signaling pathway, which reacts upon changes in the osmotic conditions, there was a clear difference between *Aspergillus* and *Penicillium* (Stoll et al. 2013). In the case of *P. verrucosum* the ochratoxin biosynthesis is regulated by the HOG signal cascade (Dihazi et al. 2004; Macheleidt et al. 2016). At conditions were ochratoxin is produced, e.g. at higher NaCl conditions (above 3%), HOG is phosphorylated and thereby active (Stoll et al. 2013) which is directly coupled to the shift from citrinin to ochratoxin biosynthesis by *P. verrucosum*. These results clearly indicate that ochratoxin biosynthesis in *Penicillium* spp. is regulated by HOG. This is in sharp contrast to the situation in *Aspergillus*. In fact, the results demonstrated that ochratoxin A biosynthesis in *Aspergillus* does not seem to be HOG regulated, which seems to be the main reason why *Aspergillus* is not able to produce ochratoxin under high NaCl conditions (Stoll et al. 2013). *P. verrucosum* almost showed no growth retardation at higher NaCl concentrations, indicating its good adaptation to these environments. This is not the case for *Aspergillus*. Growth of *A. carbonarius* is almost completely inhibited at NaCl concentrations above 5% suggesting that *Aspergillus* spp. are not adapted to this environment (Stoll et al. 2013). This explains the situation why *P. verrucosum* is rather responsible for the presence of ochratoxin in salted olives compared to *Aspergillus*. Apparently citrinin production under these conditions is not due to *P. verrucosum* but to *P. citrinum*. The latter also occurs in this environment and is a very consistent citrinin producer (Heperkan et al. 2006).

Because ochratoxin A carries chlorine in its molecule and is produced permanently in an oscillatory manner (Schmidt-Heydt et al. 2011) it was suggested that the production and excretion of ochratoxin might support the cell to keep the intracellular chlorine concentration at a physiological level. In fact, a mutant of *P. nordicum*, a species closely related to *P. verrucosum*, which was not able to produce ochratoxin, had a much higher intracellular Cl^- concentration and growth was inhibited at higher NaCl concentration, which was in clear contrast to the wild type.

Taken together, the regulation and mutual shifts of ochratoxin and citrinin biosynthesis at different conditions of the same habitat or in completely different environments seems to adapt the fungus to the specific habitat (Fig. 6.3b). In fact, again the conditions in the post-harvest environments even seem to support mycotoxin biosynthesis under the respective conditions.

Patulin and Citrinin Production by *Penicillium expansum*

P. expansum is one of the most important post-harvest pathogen of pomaceous fruits, especially apples (Baert et al. 2007; Puel et al. 2010; Errampalli 2014). *P. expansum* strains are known to produce the two important mycotoxins patulin and citrinin (Martins et al. 2002; Bragulat et al. 2008; Singh and Sumbali 2008; Santos et al. 2012; Ostry et al. 2013; Tannous et al. 2018). Strains of this species generally can colonize ripe apples and this colonization capacity is partly strain dependent (Neri et al. 2010). The production of both mycotoxins is also strain dependent (Abramson et al. 2009; De Clercq et al. 2016) and may affect the colonization capacity. Not only *P. expansum* strains but also the apple variety plays a role in determining the amount of toxins produced (Salomao et al. 2009). Usually representative amounts of patulin can be analytically determined in molded apples. In contrast, the amount of citrinin in apples is always much lower if detectable at all (Singh and Sumbali 2008; Watanabe 2008). There are some contradictory results about the role of patulin during the post-harvest colonization of apples by *P. expansum*. Sanzani et al. (2012) inactivated the *pat*K gene of the patulin biosynthesis gene cluster of *P. expansum*. The *pat*K gene codes for the patulin polyketide synthase, the key enzyme of the patulin biosynthetic pathway. The resulting transformants still produced some patulin but showed a reduced capacity to colonize apples. For that reason, the authors concluded that patulin serves as a pathogenicity factor supporting the colonization of apples. This could not be confirmed by Ballester et al. (2015). These authors inactivated the *pat*K gene too but did not find a reduced colonization capacity on apples compared to the wild type. Snini et al. (2016) could clarify this discrepancy in that they found, that the role of patulin as a colonization factor is dependent on the apple variety. The fact that patulin supports colonization of apples, especially during the post-harvest phase, points to the possibility that the natural food habitat itself supports patulin biosynthesis. That this is indeed the case was shown by Touhami et al. (2018). A *P. expansum* strain grown on laboratory medium produced less patulin than after growth on apple matrices, albeit the former medium was a rich medium, which usually supports mycotoxin biosynthesis very well. The apple environment as well as the fungus itself changes, through its metabolic activity, the local environmental conditions in a way that higher amounts of patulin are produced. This is achieved by the fungus by actively lowering the pH of the surrounding apple matrix. For this purpose, excess of gluconic acid is produced in this environment, which in turn activates patulin biosynthesis (Barad et al. 2014). According to these authors, the production of gluconic acid reduces ambient pH, which in turn activates the *pac*C gene, which thereof, as a regulatory factor, activates patulin biosynthesis.

In contrast to the patulin biosynthesis, which obviously plays a role during the colonization process, not that much was known about the ecological importance of citrinin biosynthesis by *P. expansum*, especially because of the fact that the detectable citrinin in apples is always quite low compared to that of patulin (Martins et al. 2002; Pepelnjak et al. 2002). Touhami et al. (2018) analysed the importance of the citrinin biosynthesis during the colonization of apples in the post-harvest phase in detail. They analysed the production kinetics of both toxins after growth of *P. expansum* on apples and could show that patulin is always produced immediately at beginning of the growth phase of the fungus, whereas citrinin could only be detected after storage of the infected apples for about 10 days. Interestingly the expression of the polyketide synthase genes of both toxins, e.g. the *pat*K gene of the patulin biosynthetic pathway and the *pks*CT gene of the citrinin biosynthesis pathway, were expressed very early during growth on the apple matrix. That suggests also an early production of citrinin. As described previously citrinin is regarded as an antioxidant and is especially produced at increasing oxidative conditions as well as under challenging light conditions, it seems to have protective properties. In the work of Touhami et al. (2018) it could be shown that the ROS concentration in harvested and stored apples increases after infection with *P. expansum* e. g. the apples reacts against the attack by the fungus. As a consequence, *P. expansum* could increases the expression of the *pks*CT gene to produce citrinin as a protectant. In case that citrinin acts as an antioxidant, the citrinin molecule scavenges radicals and the chemical structure of citrinin is derivatized and no longer detectable by standard analytical methods. For that reason, Touhami et al. (2018) analysed possible oxidation products of citrinin after treatment with an oxidant (hydrogen peroxide) in an aqueous solution. The authors could show that citrinin is derivatized into 3 metabolites. Interestingly the same citrinin derivatives were found after growth of *P. verrucosum* on apple matrix, indicating that citrinin is degraded also in this environment by the increased ROS. Additional to this degrading effect citrinin binds very tightly to pectin (Touhami et al. 2018) and is no longer extractable after binding. This effect of course also withdraws citrinin from analytical detection. Taken together, despite the fact that citrinin cannot be analytically detected in the early growth phase of *P. expansum* on apples, the early expression of the *pks*CT gene suggests an earlier production. Interestingly the presence of analytically detectable amounts of citrinin after growth on apple coincides with the formation of aerial mycelium (Fig. 6.4). During the early growth phase *P. expansum* predominantly grows in the form of substrate mycelium inside the apple. Only after a storage period of about 8–10 days at 25 °C a considerable amount of aerial mycelium is produced. This mycelium also starts to sporulate vigorously at that time. So, it looks that citrinin, like patulin, is obviously produced over the whole growth phase but that analytically detectable amounts are only produced when aerial mycelium is produced. Citrinin produced by the substrate mycelium seems to be either bound to pectin or degraded by the ROS of the apple matrix.

Touhami et al. (2018) inactivated the *pks*CT gene. The resulting transformant could not produce citrinin and showed a reduced capacity to colonize apples. In fact, after inoculation in apples the transformant started to grow and lead to an initial infection, however this initial rotten area stayed very compact and did not spread to an extended area usually observed after infection with the wildtype. The addition of external citrinin however restored the colonization capacity of the transformant. These results clearly show a contribution of citrinin to the colonization of post-harvest apples. However, as Touhami et al. (2018) also showed with *Penicillium* strains which produced different amounts of citrinin, the colonization capacity and the amount of citrinin produced are not ultimately coupled. This suggests, as in the case with the other mycotoxins, that these secondary metabolites only accessorily support adaptation to a certain habitat but are not ultimately needed.

In any case, as with the regulation of ochratoxin and citrinin in the case of *P. verrucosum*, the natural habitat of *P. expansum* seem to favor the biosynthesis of

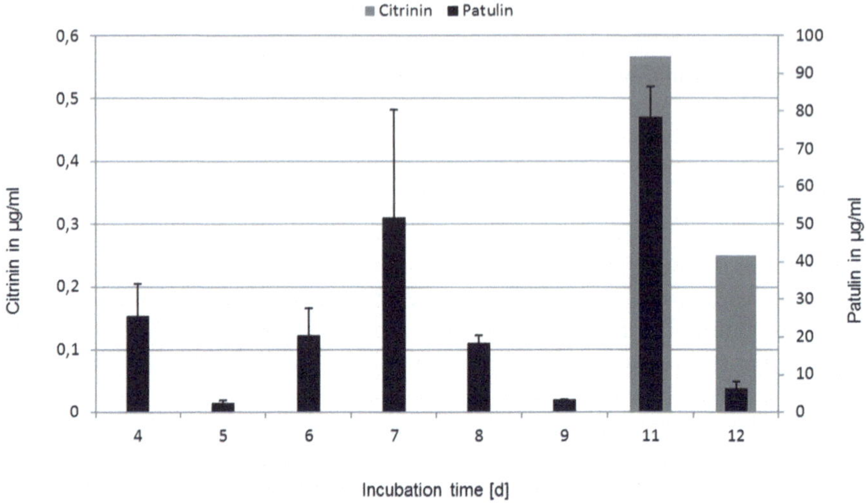

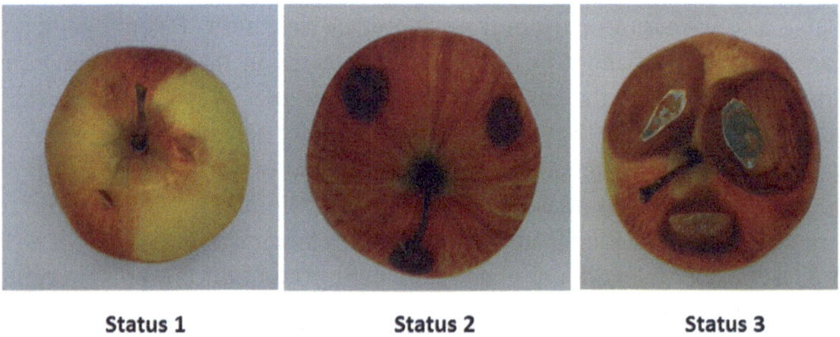

Fig. 6.4 Example of a toxin production kinetics of a strain of *P. verrucosum* grown on apples at 20 °C for certain days. Generally, patulin can be detected very early during the growth phase whereas citrinin can usually only be found after prolonged storage
Stage 1: Early phase, shortly after infection, fungus grows within the apple matrix: The *pat*K gene and the *pks*CT gene are usually significantly expressed. Patulin can be detected analytically, citrinin usually not. According to Touhami et al. (2018) citrinin is apparently be produced but immediately bound to pectin or degraded by the elevated ROS concentrations of the infected apple matrix. Stage 2: Fungus still produces only substrate mycelium, patulin is still produced, but concentration may oscillate. Still usually no citrinin can be analytically detected. Expression of both genes is reduced. Stage 3: Citrinin can be detected analytically. Aerial mycelium and spores are produced. The rotten area is in its final state

patulin and citrinin, because both toxins support the adaptation of the fungus to the somewhat challenging environments.

Alternariol Production by *A. alternata*

A. alternata is a mycotoxin producing fungus which occurs on various fruits and vegetables like tomatoes, carrots, pears, cole crops, parsley but also wheat (Logrieco et al. 2009). *A. alternata* strains can be differentiated into true pathogenic and saprophytic strains. Pathogenic strains contain a so called conditionally dispensable chromosome (CDC) which carries host specific toxins. Saprophytic strains do not contain such a chromosome however can receive it from a pathogenic strain thereby getting pathogenic (Hatta et al. 2002; Akagi et al. 2009). Important mycotoxins produced by *A. alternata* are alternariol, alternariol monomethyl ether and tenuazonic acid. The gene clusters of these mycotoxins are not located on CDC's. Tenuazonic acid is produced in high amounts which are quite independent from external parameters. This is not the case for alternariol. Graf et al. (2012) showed that the production of alternariol by *A. alternata* is more or less independent over a broad pH-range but its production is immediately ceased if the osmotic activity of the substrate is only slightly increased by NaCl. That means that alternariol biosynthesis is increased under conditions which resemble the natural post-harvest habitat of fruits and vegetables. These substrates have usually a high water activity but can have a broad pH-spectrum. *A. alternata* often infects packaged tomatoes during the distribution chain. During this infection process alternariol plays an important role as colonization factor (Graf et al. 2012). A mutant in alternariol biosynthesis was not able to colonize wounded tomatoes however, the external addition of alternariol restored the colonization capacity in a concentration dependent manner. The production of alternariol is regulated according the conditions in the natural habitat. At higher osmotic conditions, which do not indicate fruits and vegetables as natural habitat, alternariol biosynthesis is ceased through regulation of the HOG- pathway, a general MAP kinase signal transduction pathway. However, at certain pH-conditions, which may occur in the habitat of *Alternaria*, alternariol biosynthesis is regulated by *pac*C (Geisen et al. 2015). In this way the fungus ensures that alternariol, as a colonization factor, is produced when it is needed, that means at conditions with high water activity but in a broad pH range which represents fruits and vegetables.

Again, the conditions in the natural habitat induce the synthesis of the mycotoxin alternariol.

Conclusion

In the literature different reasons are discussed why certain fungi produce mycotoxins. Ehrlich (2006) for example mentioned as possible biological reasons for the biosynthesis of mycotoxins the response of fungi to stress, the protection from UV damage, their biosynthesis as by-products of primary metabolism, the increase of spore yields or protection against insect predators. Especially the last aspect is one important point in terms of the post-harvest environment (Hodge 1996; Rohlfs et al. 2007). In their review Macheleidt et al. (2016) also discussed some possible ecological roles for the biosynthesis of mycotoxins. Among them they mentioned that they are produced to interact with other microorganisms in the same habitat. Indeed, some of the mycotoxins like patulin or citrinin have antibacterial properties (Boukouvalas and Maltais 1995; Mazumder et al. 2001) and enable the fungus to combat against certain microorganisms which may colonize the same habitat. Macheleidt et al. (2016) also mentioned the possibility that mycotoxins act as pathogenicity or colonization factors. The first term is more suited for true pathogens (field fungi) the last term fits for factors supporting the colonization of plant parts after harvest. In the current article it was discussed that cyclopiazonic acid, patulin, citrinin and alternariol play such a role as colonization factors, whereas aflatoxin and ochratoxin seem to act as adaptation factors. Generally, the production of mycotoxins is correlated to increased stress conditions, especially oxidative stress conditions. These facts strengthen the view that the plant environment, also during the post-harvest state, may be inducing for mycotoxin biosynthesis at certain conditions. However, mycotoxins seem not to play a central but an accessory role for the colonization of the habitat (Touhami et al. 2018). That is obviously one of the reasons why mutants in mycotoxin biosynthesis still can grow on laboratory media and even in the natural environment. According to Roze et al. (2011) mycotoxin production is not essential at least for short term survival. The syntheses of certain mycotoxins, whose production are elevated under stress conditions, seem to support the adaptation to the environment but are not ultimately necessary for growth in this respective environment. In their review Barad et al. (2016) clearly describe, according to the literature available, that patulin acts only as a pathogenicity factor under certain conditions e. g. in certain apple varieties, whereas after growth on other varieties patulin does not seem to have any effect. However, because they seem to support adaptation, albeit only accessory, on the long run mycotoxin producing strains are apparently more assertive than mycotoxin non-producing strains. According to Fox and Howlett (2008) the production of a certain secondary metabolite is not essential but only supports the fungus to survive in the respective ecological niche. Taken together, this fact means that this ecological niche favors the biosynthesis of certain mycotoxins, as shown by the examples described above.

References

Abramson D, Lombaerd GA, Clear RM, Sholberg P, Trelka R, Rosin E (2009) Production of patulin and citrinin by *Penicillium expansum* from British Columbia (Canada) apples. Mycotox Res 25:85–88

Abriouel H, Benomar N, Lucas R, Gálvez A (2011) Culture-independent study of the diversity of microbial populations in brines during fermentation of naturally-fermented Aloreña green table olives. Int J Food Microbiol 144:487–496

Akagi Y, Akamatsu H, Otani H, Kodama M (2009) Horizontal chromosome transfer, a mechanism for the evolution and differentiation of a plant-pathogenic fungus. Eukaryot Cell 8:1732–1738

Alkan N, Fortes AM (2015) Insights into molecular and metabolic events associated with fruit response to post-harvest fungal pathogens. Front Plant Sci 6:889

Asghari M, Aghdam MS (2010) Impact of salicylic acid on post-harvest physiology of horticultural crops. Trends Food Sci Technol 21:502–509

Aver'yanov AA, Lapikova VP, Lebrun MH (2007) Tenuazonic acid, toxin of rice blast fungus, induces disease resistance and reactive oxygen production in plants. Russ J Plant Physiol 54:749–754

Aver'yanov AA, Belozerskaya TA, Gessler NN (2012) Fungus development and reactive oxygen: phytopathological aspects. In: Witzany G (ed) Biocommunication of Fungi. Springer, Dordrecht, pp 261–271

Baert K, Devlieghere F, Flyps H, Oosterlinck M, Ahmed MM, Verlinden B, Nicolai B, Debevere J, De Meulenaer B (2007) Influence of storage conditions of apples on growth and patulin production by *Penicllium expansum*. Int J Food Microbiol 119:170–181

Ballester AR, Marcet-Houben M, Levin E, Sela N, Selma-Lázaro C, Carmona L, Wisniewski M, Droby S, Gonzáles-Candelas L, Gabaldón T (2015) Genome, transcriptome, and functional analyses of *Penicillium expansum* provide new insights into secondary metabolism and pathogenicity. MPMI 28:232–248

Barad S, Horowitz SB, Kobiler I, Sherman A, Prusky D (2014) Accumulation of the mycotoxin patulin in the presence of gluconic acid contributes to pathogenicity of *Penicillium expansum*. MPMI 27:66–77

Barad S, Sionov E, Prusky D (2016) Role of patulin in post-harvest diseases. Fungal Biol Rev 30:24–32

Bavaro SL, Susca A, Frisvad J, Tufariello M, Chytiri A, Perrone G, Mita G, Logrieco A, Bleve G (2017) Isolation, characterization, and selection of molds associated to fermented black table olives. Front Microbiol 8. Article 1356

Bhatnagar D, Yu J, Ehrlich KC (2002) Toxins of filamentous fungi. In: Breitenbach M, Crameri R, Lehrer SB (eds) Fungal allergy and pathogenicity. Karger, Basel, pp 167–206

Boukouvalas J, Maltais F (1995) An efficient synthesis of the antibiotic patulin. Tetrahedron Lett 36:7175–7176

Bragulat MR, Abarca ML, Cabanes FJ (2008) Low occurrence of patulin- and citrinin-producing species isolated from grapes. Lett Appl Microbiol 47:286–289

Cabanas R, Bragulat MR, Abarca ML, Castella G, Cabanes FJ (2008) Occurrence of *Penicillium verrucosum* in retail wheat flours from the Spanish market. Food Microbiol 25:642–647

Chalivendra SC, DeRobertis C, Chang PK, Damann KE (2017) Cyclopiazonic acid is a pathogenicity factor for *Aspergillus flavus* and a promising target for screening germplasm for ear rot resistance. MPMI 30:361–373

Chang PK, Ehrlich KC (2011) Cyclopiazonic acid biosynthesis by *Aspergillus flavus*. Toxin Rev 30:71–78

Chang PK, Horn BW, Dorner JW (2009a) Clustered genes involved in cyclopiazonic acid production are next to the aflatoxin biosynthesis gene cluster in *Aspergillus flavus*. Fungal Genet Biol 46:176–182

Chang PK, Ehrlich KC, Fujii I (2009b) Cycopiazonic acid biosynthesis of *Aspergillus flavus* and *Aspergillus oryzae*. Toxins 1:74–99

De Clercq N, Vlaemynck G, Van Pamel E, Colman D, Heydricks M, Van Hove F, De Meulenaer B, Devlieghere F, Van Coillie E (2016) Patulin production by *Penicillium expansum* isolated from apples during different steps of long-term storage. World Mycotoxin J 9:379–388

Desmond OJ, Manners JM, Stephens AE, Maclean DJ, Schenk PM, Gardiner DM, Munn AL, Kazan K (2008) The *Fusarium* mycotoxin deoxynivalenol elicits hydrogenperoxide production, programmed cell death and defense responses in wheat. Mol Plant Pathol 9:435–445

Dhungana B, Ali S, Byamukama E, Krishnan P, Caffe-Treml M (2018) Incidence of *Penicillium verrucosum* in grain samples from oat varieties commonly grown in South Dakota. J Food Prot 81:898–902

Dihazi H, Kessler R, Eschrich K (2004) High osmolarity glycerol (HOG) pathway-induced phosphorylation and activation of 6-phosphofructo-2-kinase are essential for glycerol accumulation and yeast cell proliferation under hyperosmotic stress. J Biol Chem 279:23961–23968

Ehrlich KC (2006) Evolution of the aflatoxin gene cluster. Mycotoxin Res 22:9–15

El Adlouni C, Tozlovanu M, Naman F, Faid M, Pfohl-Leszkowicz A (2006) Preliminary data on the presence of mycotoxins (ochratoxin A. citrinin and aflatoxin B1) in black table olives "Greek style" of Moroccan origin. Mol Nutr Food Res 50:507–512

El-Esawi M, Arthaut LD, Jourdan N, d'Harlingue A, Link J, Martino CF, Ahmad M (2017) Blue-light induced biosynthesis of ROS contributes to the signaling mechanism of *Arabidopsis* cryptochrome. Sci Rep 7:13875

Elmholt S (2003) Ecology of ochratoxin a producing *Penicillium verrucosum*: occurrence in field soil and grain with special attention to farming system and on-farm drying practices. Biol Agric Hortic 20:311–337

Elmholt S, Hestbjerg H (1999) Field ecology of the ochratoxin A-producing *Penicillium verrucosum*: survival and resource colonization in soil. Mycopathologia 147:67–81

Errampalli D (2014) *Penicillium expansum* (Blue Mold). In: Bautista-Baños S (ed) Postharvest decay – control strategies, academic press, London, pp 189–231

Fernández G, Fernández Díez MJ, Adams MR (1997) Table olives, production and processing. Chapman and Hall, London

Ferracane R, Tafuri A, Logrieco A, Galvano F, Balzano D, Ritieni A (2007) Simultaneous determination of aflatoxin B1 and ochratoxin A and their natural occurrence in Mediterranean virgin oil. Food Addit Contam 24:173–180

Finoli C, Vecchio A, Galli A, Franzetti L (1999) Production of cyclopiazonic acid by molds isolated from Taleggio cheese. J Food Prot 62:1198–1202

Fox EM, Howlett BJ (2008) Secondary metabolism: regulation and role in fungal biology. Curr Opin Microbiol 11:481–487

Frisvad JC, Lund F, Elmholt S (2005) Ochratoxin A producing *Penicillium verrucosum* isolates from cereals reveal large AFLP fingerprinting variability. J Appl Microbiol 98:684–692

Gallo A, Solfrizzo M, Epifani F, Panzarini G, Perrone G (2016) Effect of temperature and water activity on gene expression and aflatoxin biosynthesis in *Aspergillus flavus* on almond medium. Int J Food Microbiol 217:162–169

Geisen R, Graf E, Schmidt-Heydt M (2015) HogA and PacC regulated alternariol biosynthesis by *Alternaria alternata* is important for successful substrate colonization. In: Ippolito A, Wisniewski M, Droby S (eds) Proceedings of the III international symposium on Postharvest Pathology: Using Science to Increase Food Availability, Acta Horticulturae 1144, pp 141–148

Ghitakou S, Koutras K, Kanellou E, Markaki P (2006) Study of aflatoxin B_1 and ochratoxin A production by natural microflora and *Aspergillus parasiticus* in black and green olives of Greek origin. Food Microbiol 23:612–621

Gomez AAM, Queiroz MV, Pereira OL (2015) Mycofumigation for the biocontrol of post-harvest diseases in fruits and vegetables: a review. Austin J Biotechnol Bioeng 2:1051

Graf E, Schmidt-Heydt M, Geisen R (2012) HOG MAP kinase regulation of alternariol biosynthesis in *Alternaria alternata* is important for substrate colonization. Int J Food Microbiol 157:353–359

Hatta R, Ito K, Hosaki Y, Tanaka T, Tanaka A, Yamamoto M, Akimitsu K, Tsuge T (2002) A conditionally dispensable chromosome controls host-specific pathogenicity in the fungal plant pathogen *Alternaria alternata*. Genetics 161:59–70

Heider EM, Harper JK, Grant DM, Hoffmann A, Dugan F, Tomer DP, O'Neil KO (2006) Exploring unusual antioxidant activity in a benzoic acid derivative: a proposed mechanism for citrinin. Tetrahedron 62:1199–1208

Heperkan D, Meric BE, Sismanoglu G, Dalkilic G, Güler FK (2006) Mycobiota, mycotoxigenic fungi and citrinin production in black olives. In: Hocking A, Samson R, Pitt JI, Thrane C (eds) Advances in food mycology. Springer, New York, pp 203–210

Heperkan D, Dazkir GS, Kansu DZ, Güler FK (2009) Influence of temperature on citrinin accumulation by *Penicillium citrinum* and *Penicillium verrucosum* in black table olives. Toxin Rev 28:180–186

Hodge S (1996) The relationship between *Drosophila* occurrence and mould abundance on rotting fruit. Br J Entomol Nat Hist 9:87–91

Hong SY, Roze LV, Linz J (2013) Oxidative stress-related transcriptional factors in the regulation of secondary metabolism. Toxins 5:683–702

Hurtado A, Reguant C, Bordons A, Rozès N (2009) Influence of fruit ripeness and salt concentration on the microbial processing of Arbequina table olives. Food Microbiol 26:827–833

Hymery N, Vasseur V, Coton M, Mournier J, Jany JL, Barbier G, Coton E (2014) Filamentous fungi and mycotoxins in cheese: a review. Compr Rev Food Sci Food Saf 13:437–456

Iamanaka BT, Castle de Menezes H, Vicente E, Leite RSF, Taniwaki MH (2007) Aflatoxigenic fungi and aflatoxins occurrence in sultanas and dried figs commercialized in Brazil. Food Control 18:454–457

Ishikawa Y, Morimoto K, Hamasaki T (1988) Citrinin as a potent antioxidant compound from *Penicillium* species. J Jpn Oil Chem Soc 37:51–53

Jayashree T, Subramanyam C (2000) Oxidative stress as a prerequisite for aflatoxin production by *Aspergillus parasiticus*. Free Radic Biol Med 29:981–985

Lansden JA, Davison JI (1983) Occurrence of cyclopiazonic acid in peanuts. Appl Environ Microbiol 45:766–769

Le Bars J (1979) Cyclopiazonic acid production by *Penicillium camemberti* Thom and natural occurrence of this mycotoxin in cheese. Appl Environ Microbiol 38:1052–1055

Limay-Rios V, Miller JD, Schaafsma AW (2017) Occurrence of *Penicillium verrucosum*, ochratoxin A, ochratoxin B and citrinin in on-farm stored winter wheat from the Canadian Great Lakes Region. PLoS One 12:e0181239

Lindblad M, Johnsson P, Jonsson N, Lindqvist R, Olsen M (2004) Predicting non-compliant levels of ochratoxin A in cereal grain from *Penicillium verrucosum* counts. J Appl Microbiol 97:609–616

Liu J, Sui Y, Wisniewski M, Xie Z, Liu Y, You Y, Zhang X, Sun Z, Li Y, Wang Q (2018) The impact of the post-harvest environment on the viability and virulence of decay fungi. Crit Rev Food Sci Nutr 58:1681–1687

Logrieco A, Moretti A, Solfrizzo M (2009) *Alternaria* toxins and plant diseases: an overview of origin, occurrence and risks. World Mycotoxin J 2:129–140

Lund F, Frisvad JC (2003) *Penicillium verrucosum* in wheat and barley indicates presence of ochratoxin A. J Appl Microbiol 95:1117–1123

Macheleidt J, Mattern DJ, Fischer J, Netzker T, Weber J, Schroeckh V, Valiante V, Brakhage AA (2016) Regulation and role of fungal secondary metabolites. Annu Rev Genet 50:371–392

Magan N, Hope R, Cairns V, Aldred D (2003) Post-harvest fungal ecology: impact of fungal growth and mycotoxin accumulation in stored grain. Eur J Plant Pathol 109:723–730

Mahoney N, Molyneux RJ, Kim JH, Campbell BC, Waiss AC, Hagerman AE (2010) Aflatoxigenesis induced in *Aspergillus flavus* by oxidative stress and reduction by phenolic antioxidants from tree nuts. World Mycotoxin J 3:49–57

Maor U, Sadhasivam S, Zakin V, Prusky D, Sionov E (2017) The effect of ambient pH modulation on ochratoxin A accumulation by *Aspergillus carbonarius*. World Mycotoxin J 10:339–348

Martins ML, Gimeno A, Martins HM, Bernado F (2002) Co-occurrence of patulin and citrinin in Portuguese apples with rotten spots. Food Addit Contam 19:568–574

Massi FP, Vieira MLC, Sartori D, Penha RES, de Freitas Munhoz C, Ferreira JM, Iamanaka BT, Taniwaki MH, Frisvad J, Fungaro MHP (2014) Brazil nuts are subject to infection with B and G aflatoxin-producing fungus, *Aspergillus pseudonomius*. Int J Food Microbiol 186:14–21

Mazumder PM, Mazumder R, Mazumder A, Sasmal DS (2001) Antimicrobial activity of the mycotoxin citrinin obtained from the fungus *Penicillium citrinum*. Anc Sci Life 21:191–197

Mittler R (2012) Oxidative stress, antioxidants and stress tolerance. Trends Plant Sci 7:405–410

Neri F, Donati I, Veronesi F, Mazzoni D, Mari M (2010) Evaluation of *Penicillium expansum* isolates for aggressiveness, growth and patulin accumulation in usual and less common fruit hosts. Int J Food Microbiol 143:109–117

Ostry V, Malir F, Ruprich J (2013) Producers and important dietary sources of ochratoxin A and citrinin. Toxins 5:1574–1586

Oviedo MS, Ramirez ML, Barros GG, Chulze SN (2010) Impact of water activity and temperature on growth and alternariol and alternariol monomethyl ether production of *Alternaria alternata* isolated from soybean. J Food Prot 73:336–343

Palencia ER, Palencia ER, Hinton DM, Bacon CW (2010) The black *Aspergillus* species of maize and peanuts and their potential for mycotoxin production. Toxins 2:399–416

Papachristou A, Markaki P (2004) Determination of ochratoxin A in virgin olive oils of Greek origin by immunoaffinity column clean-up and high-performance liquid chromatography. Food Addit Contam 21:85–92

Pepelnjak S, Segvic M, Ozegovic L (2002) Citrinotoxinogenicity of *Penicillium* spp. isolated from decaying apples. Braz J Microbiol 33:134–137

Pinto VF, Patriarca A, Locani O, Vaamonde G (2001) Natural co-occurrence of aflatoxin and cyclopiazonic acid in peanuts grown in Argentina. Food Addit Contam 18:1017–1020

Pleadin J, Frece J, Kudumija N, Petrović D, Višnja V, Zadravec M, Šrivanko M, Perković I, Markov K (2016) Citrinin in cereals and feedstuffs coming from Croatia and Bosnia & Herzegovina. Food Addit Contam 9:268–274

Puel O, Galtier P, Oswald IP (2010) Biosynthesis and toxicological effects of patulin. Toxins 2:613–631

Reddy VK, Reddy SM (1992) Cyclopiazonic acid production by *Penicillium griseofulvum* in relation to different cultivars of maize. World J Microbiol Biotechnol 8:208–209

Resnik SL, González HHL, Pacin AM, Viora M, Caballero GM, Gros EG (1996) Cyclopiazonic acid and aflatoxins production by *Aspergillus flavus* isolated from Argentinian corn. Mycotoxin Res 12:61–66

Reverberi M, Fabbri AA, Zjalic S, Ricelli A, Punelli F, Fanelli C (2005) Antioxidant enzymes stimulation in *Aspergillus parasiticus* by *Lentinula edodes* inhibits aflatoxin production. Appl Microbiol Biotechnol 69:207–215

Reverberi M, Zjalic S, Ricelli A, Fabbri AA, Fanelli C (2006) Oxidant/antioxidant balance in *Aspergillus parasiticus* affects aflatoxin biosynthesis. Mycotoxin Res 22:39–47

Reverberi M, Ricelli A, Zjalic S, Fabbri AA, Fanelli C (2010) Natural functions of mycotoxins and control of their biosynthesis in fungi. Appl Microbiol Biotechnol 87:899–911

Reverberi M, Gazetti K, Punelli F, Scarpari M, Zjalik S, Ricelli A, Fabbri AA, Fanelli C (2012) Ao*yap*1 regulates OTA synthesis by controlling cell redox balance in *Aspergillus ochraceus*. Appl Microbiol Biotechnol 95:1293–1304

Reverberi M, Punelli M, Scala V, Scarpari M, Uva P, Mentzen WI, Dolezal AL, Woloshuk C, Pinzari F, Fabbri AA, Fanelli C, Payne GA (2013) Genotypic and phenotypic versatility of *Aspergillus flavus* during maize exploitation. PLoS One 8:e68735

Rohlfs M, Albert M, Keller NP, Kempken F (2007) Secondary chemicals protect mould from fungivory. Biol Lett 3:523–525

Roze LV, Chanda A, Linz JE (2011) Compartmentalization and molecular traffic in secondary metabolism: a new understanding of established cellular processes. Fungal Genet Biol 48:35–48

Roze LV, Hong SY, Linz JE (2013) Aflatoxin biosynthesis: current frontiers. Annu Rev Food Sci Technol 4:293–311

Salomao BCM, Aragao GMF, Churey JJ, Padilla-Zakour OI, Worobo RW (2009) Influence of storage temperature and apple variety on patulin production by *Penicillium expansum*. J Food Prot 72:1030–1036

Samane S, Tantaoui-Elaraki A, Essadaoui M (1991) Mycoflora of Moroccan "Greek style" black olives. II Toxigenesis. Microbiol Aliment Nutr 9:335–352

Santos IM, Abrunhosa L, Venancio A, Lima N (2012) The effect of culture preservation techniques on patulin and citrinin production by *Penicillium expansum* link. Lett Appl Microbiol 35:272–275

Sanzani SM, Reverberi M, Punelli M, Ippolito A, Fanelli C (2012) Study on the role of patulin on pathogenicity and virulence of *Penicillium expansum*. Int J Food Microbiol 153:323–331

Sanzani SM, Reverberi M, Geisen R (2016) Mycotoxins in harvested fruits and vegetables: insights in producing fungi, biological role, conducive conditions, and tools to manage postharvest contamination. Postharvest Biol Technol 122:95–105

Schmidt-Heydt M, Magan N, Geisen R (2008) Stress induction of mycotoxin biosynthesis genes b abiotic factors. FEMS Microbiol Lett 284:142–149

Schmidt-Heydt M, Bode H, Raupp F, Geisen R (2010) Influence of light on ochratoxin biosynthesis by *Penicillium*. Mycotox Res 26:1–8

Schmidt-Heydt M, Rüfer C, Raupp F, Bruchmann A, Perrone G, Geisen R (2011) Influence of light on food relevant fungi with emphasis on ochratoxin producing species. Int J Food Microbiol 145:229–237

Schmidt-Heydt M, Cramer B, Graf I, Lerch S, Humpf HU, Geisen R (2012a) Wavelength-dependent degradation of ochratoxin and citrinin by light *in vitro* and *in vivo* and its implications on *Penicillium*. Toxins 4:1535–1551

Schmidt-Heydt M, Graf E, Stoll D, Geisen R (2012b) The biosynthesis of ochratoxin A by *Penicillium* as one mechanism for adaptation to NaCl rich foods. Food Microbiol 29:233–241

Schmidt-Heydt M, Stoll D, Schütz P, Geisen R (2015) Oxidative stress induces the biosynthesis of citrinin by *Penicillium verrucosum* at the expense of ochratoxin. Int J Food Microbiol 192:1–6

Singh Y, Sumbali G (2008) Patulin and citrinin production in apple and pear fruits by *Penicillium expansum* strains collected from Jammu markets. Indian Phytopathol 61:471–477

Snini SP, Tannous J, Heuillard P, Bailly S, Lippi Y, Zehraoui E, Barreau C, Oswald IP, Puel O (2016) Patulin is a cultivar-dependent aggressiveness factor favouring the colonization of apples by *Penicillium expansum*. Mol Plant Pathol 17:920–930

Stoll D, Schmidt-Heydt M, Geisen R (2013) Differences in the regulation of ochratoxin A by the HOG pathway in *Penicillium* and *Aspergillus* in response to high osmolar environments. Toxins 5:1282–1298

Størmer FC, Sandven P, Huitfeldt S, Eduard W, Skogstad A (1998) Does the mycotoxin citrinin function as a sun protectant in conidia from *Penicillium verrucosum*? Mycopathologia 142:43–47

Tannous J, Keller NP, Atoui A, El Khoury A, Lteif R, Oswald IP, Puel O (2018) Secondary metabolism in *Penicillium expansum*: emphasis on recent advances in patulin research. Crit Rev Food Sci Nutr 58:2082–2098

Tittlemier S, Roscoe M, Blagden R, Kobialka C (2014) Occurrence of ochratoxin A in Canadian wheat shipments, 2010–2012. Food Addit Contam 5:910–916

Tokuşoğlu Ö, Bozoğlu F (2010) Citrinin risk in black and green table olives: simultaneous determination with ochratoxin-A by optimized extraction and IAC-HPLC-FD. Ital J Food Sci 22:284–291

Touhami N, Soukup ST, Schmidt-Heydt M, Kulling S, Geisen R (2018) Citrinin as an accessory establishment factor of *P. expansum* for the colonization of apples. Int J Food Microbiol 266:224–233

Trzaska WJ, Wrigley HE, Thwaite JE, May RC (2016) Species-specific antifungal activity of blue light. Sci Rep 7:1–7

Vrabcheva T, Usleber E, Dietrich R, Märtlbauer E (2000) Co-occurrence of ochratoxin A and citrinin in cereals from Bulgarian villages with a history of Balkan endemic nephropathy. J Agric Food Chem 48:2483–2488

Watanabe M (2008) Production of mycotoxins by *Penicillium expansum* inoculated into apples. J Food Prot 71:1714–1719

Wawrzyniak J, Waśkiewicz A (2014) Ochratoxin A and citrinin production by *Penicillium verrucosum* on cereal solid substrates. Food Addit Contam 31:139–148

Yu J, Bhatnagar D, Cleveland TE (2004) Completed sequence of aflatoxin pathway gene cluster in *Aspergillus parasiticus*. FEBS Lett 564:126–130

Zhao X, Zhi Q, Li JY, Keller NP, Zhu-Mei H (2018) The antioxidant gallic acid inhibits aflatoxin formation in *Aspergillus flavus* by modulating transcription factors FarB and CreA. Toxins 10:270–287

Chapter 7
Innovative Strategies for the Management of *Aspergillus* spp. and *Penicillium* spp. on Nuts

Davide Spadaro, Mauro Fontana, Simona Prencipe, Silvia Valente, Edoardo Piombo, and Maria Lodovica Gullino

Abstract A significant part of the microbiota of nuts is represented by *Aspergillus* spp., whereas *A. flavus* and *A. parasiticus* are the main aflatoxigenic species of *Aspergillus* section *Flavi*. Nut production can be contaminated by mould development and by the mycotoxins. Some mycotoxins, such as aflatoxins, are regulated at European level and in most extra-EU countries. Besides, several species of *Penicillium* often contaminate nuts during the supply chain, from harvest to storage. Beyond the economic damage, *Penicillium* spp. can also produce toxic compounds currently unregulated by European laws, such as penitrem A, roquefortine C, patulin, andrastin A, cyclopenin, cyclopenol and chaetoglobosin A. It is therefore necessary to develop harvesting methods, storage conditions, processing techniques and detoxification protocols for the management of fungal growth and mycotoxin contamination. The efficacy, the benefits and the drawbacks of drying technologies, roasting and cold plasma in reducing mycotoxin contamination are described. The information gained about the species of *Aspergillus* and *Penicillium* and their mycotoxins are useful for the stakeholders of the nut production chain in order to develop appropriate practice for the field and postharvest phases, useful to obtain healthy and safe products.

Keywords Aflatoxins · Chestnut · Hazelnut · Mycotoxins · Ochratoxin A

D. Spadaro (✉) · S. Prencipe · S. Valente · E. Piombo · M. L. Gullino
Centre of Competence for the Innovation in the Agroenvironmental Sector (AGROINNOVA) and Department of Agricultural, Forest and Food Sciences, University of Torino, Grugliasco, TO, Italy
e-mail: davide.spadaro@unito.it

M. Fontana
Soremartec Italia S.R.L., Alba, CN, Italy

D. Spadaro et al. (eds.), *Postharvest Pathology*, Plant Pathology in the 21st Century 11, https://doi.org/10.1007/978-3-030-56530-5_7

Introduction

Nuts are mainly consumed as fresh or roasted products, but also as derivatives, i.e. oils, ingredients for sauces preparation or nut creams. The commercially most important nuts are walnuts (*Juglans regia*), almonds (*Prunus amigdalis*), chestnuts (*Castanea sativa*), pistachios (*Pistacia vera*), hazelnuts (*Corylus avellana*), peanuts (*Arachis hypogea*), pine nuts (*Pinus pinea*) and cashews (*Anacardium occidentale*) (Ros 2010). Although various kind of nuts have different nutritional characteristics, they are generally considered highly nutritious, mainly due to the richness in polyunsaturated fatty acids (PUFA), proteins, fibres, vitamins and minerals (Ros 2010).

Benefits for human health linked to the consumption of nuts are widely demonstrated, and their introduction in guidelines for healthy nutrition has led to an increase in demand and consumption (Alasalvar and Shahidi 2008). World production of walnuts was 3.7 million t in 2018, with China (43.3%), the USA (16.7%) and Iran (11.2%) as main producers, the production of almonds was 3.2 million t, with the USA (58.8%) and Spain (10.7%) as main producers, chestnut production (2.4 million t) was mostly dominated by China (83.5%), the production of pistachios (1.4 million t) was concentrated in Iran (40.0%), the USA (32.5%) and Turkey (17.4%), and the production of hazelnuts was 860,000 t, with Turkey (59.7%) and Italy (15.4%) as main producers (FAOSTAT 2020).

Due to the importance of nuts both for the economic point of view and for human consumption, it is important to reduce product postharvest losses and to promote their safety and quality.

Postharvest Diseases of Nuts

Postharvest losses of nuts can be caused by insects and spoilage fungi in orchard and postharvest. Nut microbiota comprises many filamentous fungi, which may circulate as airborne inoculum in the field or during harvest and processing and may reach nuts and remain in a latent phase for a certain period. Physical, chemical and biological factors can favour fungal colonization and/or mycotoxin production. For instance, insect damages can expose kernels to fungal spores present on the surface of nuts, and this is considered the main factor leading to fungal contamination in pre and postharvest (Campbel et al. 2003). Moreover, mechanical damages during harvest, such as shell cracking, or atmospheric events, such as rain and frost, can favour fungal contamination. For example, intense rainfall during harvest can favour a higher incidence of *A. flavus* in the field (Prencipe et al. 2018a).

As reported by Abrar et al. (2012), favourable conditions to fungal development can occur during storage, washing and drying phases. In particular, storage conditions, such as temperature, relative humidity, ventilation and time in warehouse, should be monitored (Abdel-Gawad and Zhori 1993).

During postharvest, nuts should be stored in dry and clean environment, under low water activity conditions in order to hamper the growth of the fungal inoculum. However, even temporary exposure to favourable conditions may result in fungal growth and/or mycotoxin formation, since the fungal inoculum is present (Bayman et al. 2002).

Optimal conditions for fungal growth do not correspond to those for secondary metabolites production, as it is the case for aflatoxins: as reported in literature, *A. flavus* optimal growth is 30–33 °C, while aflatoxins production is inhibited over 42 °C. Furthermore, as demonstrated by Trenk and Harman (1970) and Liu et al. (2017), optimal parameters for aflatoxins production are different according to the host.

The implementation of good practices during processing and postharvest aims to prevent and reduce fungal contamination and to avoid mycotoxin biosynthesis. In addition, detoxification techniques are essential to ensure food safety.

Microbiota of Nuts

Fungal microbiota of nuts is mainly represented by the genera *Aspergillus*, *Penicillium, Rhizopus* and *Fusarium* (Bayman et al. 2002; Gürses 2006). The genus *Fusarium* is mainly present in the field, while the genera *Aspergillus* and *Penicillium* are predominant in postharvest (Rodrigues et al. 2012).

One of the most important parameter linked to fungal growth is the water activity (a_w). The species of *Aspergillus* and *Penicillium* are able to grow at low levels of a_w (from 0.75 to 0.85), with an *optimum* from 0.93 to 0.98, while *Fusarium* species need a higher a_w (0.85) and have an *optimum* of 0.99.

Fungal species able to colonize nuts may promote or limit the growth of other species, and both positive and negative correlations between fungal species were observed. For instance, a positive association was observed between *A. flavus*, *A. niger* and *Penicillium* spp. on pistachios, almonds and walnuts, while *Rhizopus* spp. were negatively associated with *Penicillium* spp. in walnuts and almonds (Bayman et al. 2002).

The complex interaction and antagonism between fungal isolates present on the surface has effects on the microbiota on nuts, which can vary significantly depending on nut tree, environmental and storage conditions. Nut trees located in the same region or processed and stored in the same environment are exposed to a similar fungal inoculum, but biological, physical and chemical differences among nuts influence their internal mycoflora.

Almonds and chestnuts microbiota is mainly represented by species of *Aspergillus* section *Flavi* and *Penicillium,* both in the field and in postharvest (Rodrigues et al. 2012) and these fungi are widely isolated from hazelnuts, walnuts, peanuts, pistachios and brazil nuts as well (Bayman et al. 2002; Şimşek et al. 2002; Gürses 2006; Ozay et al. 2008).

The main issue associated with *Penicillium* spp. and *Aspergillus* spp. is their ability to produce a wide range of mycotoxins. Aflatoxins represent the most dangerous mycotoxins for human health, and they are produced mainly by *A. flavus* and *A. parasiticus*. The contamination with aflatoxins is widely reported on nuts, while investigation of the occurrence of other mycotoxins is often lacking (Prencipe et al. 2018c). The co-existence of many fungal species in the microbiota of nuts may cause contamination with several toxic compounds (Bayman et al. 2002).

Contamination by *Aspergillus flavus* and Aflatoxin Legal Limits

The main food safety problem during the postharvest phase of nuts is related to the presence of aflatoxins, which are mycotoxins characterized by high toxic, mutagenic and teratogenic activity, classified by the International Agency for research on cancer (IARC) as carcinogenic for human (group 1), causing cancer of the liver.

AFB1 and AFB2 are mainly produced by *A. flavus* and *A. parasiticus*, the latter of which can also produce AFG1 and AFG2. The presence of these species is widely documented on chestnuts, walnuts, pistachios, hazelnuts and peanuts (Thomson and Mehdy 1978; Abdel-Hafez and Sabah 1993; Doster and Michailides 1994; Bayman et al. 2002; Horn 2005; Prencipe et al. 2018a). In orchards, *A. flavus* is more frequently isolated than *A. parasiticus*, because of its ability to grow on the aerial part of trees, in addition to soil (EFSA 2007).

Due to the human risk associated with consumption of aflatoxin contaminated commodities, limits for the presence of these molecules on food matrices, including nuts, are regulated. The maximum permissible levels in Europe of AFB1 and the maximum sum of aflatoxins (AFB1, AFB2, AFG1 and AFG2) in nuts are specified in the Commission (EU) Regulation No 165/2010. Depending on the matrix, different levels of aflatoxins are allowed: between 2 and 10 μg/kg for products intended for direct consumption, while from 5 μg/kg to 15 μg/kg for raw products. The European legislation on mycotoxin limits has been modified several times in the last years, with the acquisition of relevant scientific information.

Presence of *Aspergillus* Section *Flavi* in Chestnuts

Different studies investigated the presence of *Aspergillus* species and their aflatoxigenic potential in commercial chestnut products (Wells and Payne 1975; Rodrigues et al. 2013). The most abundant species found in chestnut belong to the section *Flavi* of *Aspergillus* (Abdel-Gawad and Zohri 1993; Overy et al. 2003; Rodrigues et al. 2013). Sorting and roasting are useful practices to reduce the presence of *Aspergillus* section *Flavi*, but current practices are not able to completely eliminate these species and reduce aflatoxin contamination. According to Bertuzzi and collaborators (2015), the high presence of aflatoxins in chestnut flour may be related to contamination in postharvest or along the chestnut flour production phases.

In a recent study (Prencipe et al. 2018a), potential aflatoxigenic species have been isolated from fresh chestnut and from the chestnut flour production chain. On fresh chestnut, the incidence of *Aspergillus* section *Flavi* section was 53%, probably favoured by the high relative humidity recorded during harvest (CAST 2003). In total, 58 strains of *Aspergillus* section *Flavi*, collected from the orchard, the chestnut flour processing and the chestnut flours, were characterized by biological, molecular and chemical assays. A multidisciplinary approach is necessary to identify the species of *Aspergillus* section *Flavi*, by using morphology, multi-locus sequencing and analysis of aflatoxin production (Rodrigues et al. 2009; Samson et al. 2014). *A. flavus* was the dominant species, followed by *A. oryzae* var. *effusus*, *A. tamarii*, *A. parasiticus* and *A. toxicarius*. The phylogenetic analysis revealed also the presence of a high intraspecific variability, as observed by Pildain et al. (2004) and Vaamonde et al. (2003). This variability was also observed through macromorphological analysis on CYA, YES and MEA media. Aflatoxins were produced *in vitro* by 19% of strains. Since the *in vitro* assays often underestimate the number of aflatoxigenic strains (Probst and Cotty 2012), an *in vivo* assay was performed, obtaining a higher percentage of aflatoxigenic strains (40%). The presence of a high number of aflatoxigenic *Aspergillus* section *Flavi* strains, isolated from the field and along all the chestnut production phases, highlights the need to implement storage conditions, focusing on temperature, humidity and drying, able to reduce the growth of these species and reduce the potential aflatoxins contamination.

Presence of *Aspergillus* Section *Flavi* in Hazelnuts

The occurrence of *A. flavus* and *A. parasiticus* on hazelnuts is well documented, despite the hard shell which represents an effective barrier for fungal contamination (Şimşek et al. 2002; Gürses 2006; Ozay et al. 2008).

Contamination can occur during the production chain: at harvest, when nuts drop into the soil, or in the processing environment due to the presence of fungal spores. Many factors influence and promote fungal growth and aflatoxins production, such as composition of hazelnuts, temperature, humidity, harvesting methods, damage due to insects or processing, time and storage conditions (Özdemir and Devres 1999; Fontana et al. 2012).

Only part of *Aspergillus* section *Flavi* found on hazelnuts is able to produce aflatoxins (Fig. 7.1). For instance, in one research performed on Turkish hazelnuts from 2002 to 2004, the percentage of aflatoxigenic species varied between 4% and 48%, though the number of hazelnut samples contaminated by *A. flavus* or *A. parasiticus* varied from 47% to 79% (Ozay et al. 2008).

Fig. 7.1 Hazelnuts covered with *Aspergillus flavus*

Generally, most isolates of *A. parasiticus* are able to produce aflatoxins, while 40–50% of *A. flavus* isolates are aflatoxigenic (Schmidt-Heydt et al. 2010; Kabak 2016).

Though the percentage can be variable, even a small amount of aflatoxigenic species can result in a significant aflatoxin contamination of hazelnuts and derived products (Aycicek et al. 2005; Gürses 2006; Baltaci et al. 2012; Prelle et al. 2012; Golge et al. 2016; Kabak 2016).

Measures to Prevent and Reduce Aflatoxins Contamination

The Food and Agriculture Organization of the United Nations (FAO) and the World Health Organization (WHO) developed a Code of Good Practice, the "Codex Alimentarius" (2010) for all nuts of commercial interest, to standardize the management practices and the control of mycotoxins, with the aim of protecting the consumers health and ensuring the accuracy of international trade. The prevention of fungal contamination starts at harvest, which should be manual in order to reduce mechanical damages and avoid contact with the ground. As reported by Rachaputi et al. (2002), a positive correlation was found between early harvest and a lower aflatoxins presence. Moreover, the use of fungicides in orchard allows a reduction of fungal infections, with a possible reduced mycotoxin contamination. The same results can be achieved by controlling insect pests and consequently by reducing nuts damages (Campbell et al. 2003). Moreover, positive effects on the reduction of aflatoxins contamination along the production chain were also demonstrated by using biological control agents, able to reduce the growth of aflatoxigenic strains (Dorner and Cole 2001; Atehnkeng et al. 2014).

After harvest, many strategies can be applied to prevent aflatoxin contamination, such as sorting, rapid drying and appropriate storage conditions (Kader 2013). Sorting is commonly adopted to remove immature and damaged nuts as well as soil and plant residues, which may serve as inoculum to infect healthy nuts. For instance, pistachio nuts with discoloured or stained shells are usually associated with aflatoxin contamination and electronic colour sorting allows to separate these kernels (Doster and Michailides 1999). Besides visual sorting, flotation can be adopted, as infected or immature nuts are usually lighter and can easily be separated from mature and healthy nuts (Karlovsky et al. 2016).

Despite separation of defected fruits, aflatoxigenic species may be present also on fruits without visible damages (Bayman et al. 2002), making further preventive measures necessary.

The most common technique to reduce fungal growth and mycotoxins production is drying. Dehulled nuts should be dried as soon as possible, preferably within 72 h from harvesting, to a moisture content corresponding to a water activity of less than 0.70 (5–8% depending on nuts), which inhibits *A. flavus* and *A. parasiticus* growth (Codex Alimentarius 2010; Kader 2013). Drying reduces the moisture content of the nut, whereas the drying parameters – temperature and time – significantly affect fungal growth and mycotoxin production.

Additionally, unstable temperatures or inefficient dehumidification in refrigerated storage can cause a rapid increase in water activity, and a consequent proliferation of filamentous fungi (Özilgen and Özdemir 2001; Fontana et al. 2012). Moreover, containers for storage and transport of nuts should be clean, dry and free of insects and fungal contaminations (Codex Alimentarius 2010).

Storage conditions are therefore essential to limit the development of mycotoxigenic fungi and the possible production of mycotoxins.

Despite all the previously described methods, in certain situations it is simply not possible to completely avoid mycotoxigenic fungi and mycotoxins, making it necessary to implement detoxification techniques (Spadaro and Garibaldi 2017). Chemical, physical, and biological approaches for the detoxification of mycotoxins have been reported in previous studies on nuts, and include strategies such as the use of ozone, cold plasma (Siciliano et al. 2016), UV or gamma irradiation (Di Stefano et al. 2014; Mao et al. 2016), roasting treatments (Siciliano et al. 2017), or microorganisms (Shantha 1999). However, most of the current methods are not practical, due to costs, time consumption, nutritional losses, or low detoxification efficiency. The effectiveness of detoxification depends upon nut species, water content, temperature, type of mycotoxin and its concentration and binding of the mycotoxin to other elements of the matrix (Kabak et al. 2006).

Effects of Drying on *A. flavus* and Aflatoxins in Chestnuts

Because of the natural composition of chestnuts, with a high water content and sugars, these fruits are particularly susceptible to mycotoxigenic fungal species (Overy et al. 2003; Attanasio et al. 2004; Rodrigues et al. 2012) and have a limited postharvest storage time. Therefore, different technologies have been studied to increase chestnut storage time preserving their chemical composition (Moreira et al. 2011; Nazzaro et al. 2011). Drying is a common practice used to limit and prevent fungal development (Jermini et al. 2006; Rodrigues et al. 2012). One of the main problems related to chestnut storage is represented by the presence of aflatoxins: some studies have shown a high incidence of contaminated chestnut flour, as well as dried and fresh chestnuts, with values up to 92% (Pietri et al. 2012; Bertuzzi et al. 2015). Therefore, this scenario underlines the problems related to the chestnut-based productions, and the need to reduce the growth of aflatoxigenic fungi to obtain products free from aflatoxins.

Many studies have been carried out to evaluate the influence of environmental parameters on the growth of *A. flavus* and on the aflatoxins production in different matrices (Marín et al. 2012; Liu et al. 2017). Considering the possible aflatoxins contamination in chestnuts, a recent study showed the effect of different drying temperatures (30, 35, 40, 45 and 50 °C) on the growth of *A. flavus*, on the aflatoxins production and on the chemical composition of chestnuts (Prencipe et al. 2018b). The results obtained showed the highest presence of *A. flavus* for treatments at 30 °C and 35 °C, while no growth was shown for the samples treated at 45 °C or 50 °C, confirming the maximum temperature tolerated by this fungus (Pitt and Hocking 2009).

Aflatoxin concentration was higher in chestnuts treated at 40 °C, where there was also a lower growth of *A. flavus*, with data similar to those reported by Lahouar et al. (2016) on sorghum. The study of Marín et al. (2012), using predictive models on pistachio, showed the same trend, with an increase in aflatoxins concentration with the increase of temperature, and no aflatoxins production at temperature above 40 °C.

The application of drying treatments on the chestnuts quality caused some changes in the total phenolic content and the antioxidant capacity: the phenolic content increased while the antioxidant activity diminished, with the increase of the temperatures applied. These results are similar to those reported for boiled or toasted chestnuts (Barros et al. 2011; Nazzaro et al. 2011; Zhu 2011).

In commercial mills, fresh chestnuts are dried and subsequently subjected to various processes, including sorting, roasting, granulating, grinding and storage (Prencipe et al. 2018a). Currently chestnuts drying occurs at 30 °C for 3–5 days until 10% of relative humidity, but these conditions should be modified by increasing the temperature to 45 °C, in order to reduce fungal growth and aflatoxins production.

Effects of Drying on *A. flavus* and Aflatoxins in Hazelnuts

Drying is commonly adopted for hazelnuts to reduce moisture content, prevent fungal contamination and increase the shelf-life of products (Ozay et al. 2008).

In Turkey, the most common technique is sun drying on concrete or grass ground, before storing hazelnuts at room temperature for up to 24 months (Turan 2018; Turan and Karaosmanoğlu 2019). Hazelnuts are commonly dried in two steps: they are dried with the husks to reach 15–20% of moisture content, then they are dehulled and dried to reach about 6%. The time required to sun dry hazelnuts is strongly dependent on weather conditions, reaching to 3–4 weeks during wet season (Baltaci et al. 2012; Turan and Karaosmanoğlu 2019), and this extended time for drying can promote the mould growth, which may grow and produce mycotoxins during storage.

The use of drying machines, to quickly reduce the moisture content, can lead to a higher safety of hazelnuts. Ozay and collaborators (2008) revealed that different kinds of drying techniques may result in different levels of aflatoxins on hazelnuts. Many researchers highlighted the influence of drying parameters on the quality of hazelnuts, such as lipid oxidation, fatty acid profile, enzymatic activity and browning (López et al. 1997a, b, c; Özdemir and Devres 1999; Turan 2018; Turan and Karaosmanoğlu 2019).

A recent study investigated the influence of drying temperature and times on the viability of *A. flavus* and its ability to produce aflatoxins on hazelnuts (Valente et al. 2020). Nuts were inoculated with *A. flavus* (Fig. 7.2) and dried to reach 6% moisture content and a_w 0.71. After drying at 30, 35 and 40 °C, *A. flavus* was able to develop. Aflatoxins were detected only at 30 °C and 35 °C, with a higher level after drying at 30 °C. After 14 days of storage, the highest *A. flavus* and aflatoxin level were detected in samples treated at 35 °C. It is essential to evaluate efficient methods of drying, both to maintain the nutritional characteristics of nuts and to ensure a reduced presence of mycotoxigenic species and mycotoxins. In hazelnuts dried at

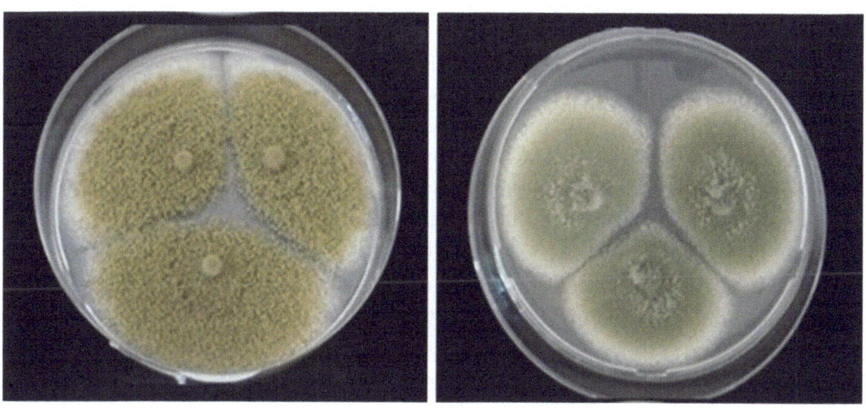

Fig. 7.2 *Aspergillus flavus* cultivated on two different media in Petri dish

45 °C, no aflatoxins were found after drying and storage, and a reduction of *A. flavus* viable conidia was observed, suggesting that 45 °C could be a good compromise between quality and safety for drying (Valente et al. 2020).

Roasting and Utilisation of Cold Plasma for Hazelnuts Detoxification

Roasting is one of the main physical means for detoxification of food matrices, but it can cause the partial or total degradation of some nutritional components. However, Siciliano et al. (2017) demonstrated that it is possible to obtain a reduction of aflatoxins without significant changes in nutritional value of hazelnuts, using an infra-red hoven to roast them at 140 °C for 40 min.

Plasma, the fourth state of matter, has been used for biological detoxification of temperature-sensitive materials. Ionization, caused by the collisions of free electrons and noble gasses, releases reactive oxygen species (positive ions, electrons, photons, excited atoms and radicals), making it possible to use plasma to disinfect materials (Laroussi 1996), inactivate microorganisms (Fridman et al. 2007; Selcuk et al. 2008) and reduce mycotoxins in food matrices (Siciliano et al. 2016). For these applications, plasma is formed in cold conditions, at atmospheric pressure, in a way that it is both cheap and able to generate active species reacting with many contaminants (Moreau et al. 2008).

The utilization of cold plasma is promising, particularly for aflatoxin detoxification on hazelnuts, where it showed a minimal impact on the organoleptic characteristics. The application of cold plasma on dehulled hazelnuts induced a detoxification from aflatoxins greater than 70% (Siciliano et al. 2016), without significantly increasing the temperature of the food matrix.

Penicillium spp. and Mycotoxins in Chestnut

Fungal contamination by genus *Penicillium* on chestnut has been studied at harvest, in storage and in commercial samples (Wells and Payne 1975; Overy et al. 2003; Rodrigues et al. 2012, 2013). The study by Bertuzzi et al. (2015) reported the contamination in different commercial chestnut products by some *Penicillium* toxins. *Penicillium* species are able to survive both at low and at high temperatures, with an optimal temperature between 25 and 30 °C (Gürses 2006). *Penicillium* species are able to cause fruits rots and are ubiquitous as contaminants in postharvest and indoor environment (Washington et al. 1997; Nielsen 2003).

Recently, the presence of *Penicillium* species has been investigated at harvest, along the chestnut production chain and indoor (Prencipe et al. 2018c): 124 strains have been characterized through biological, molecular and chemical assays, focusing on the mycotoxigenic potential and on pathogenicity.

A multidisciplinary approach, including a multi-locus genes sequencing, as well as a mycotoxins production analysis and macromorphological one, was used to identify the strains. Molecular identification was achieved by the amplification of three conserved regions. Through the amplification of the beta-tubulin gene, it was possible to identify 108 out of 124 isolates. For the 16 isolates with an ambiguous identification, the species was determined by sequencing the calmodulin gene and by macromorphological analysis on three different media. The phylogeny showed a high intraspecific variability for some species, including *P. bialowiezense*, *P. commune* and *P. glabrum*, as observed by Scott et al. (2008), Barreto et al. (2011), Houbraken et al. (2012) and Visagie et al. (2014a, b). Based on the current taxonomy reported by Houbraken and Samson (2011), Visagie et al. (2014a, b) and Houbraken et al. (2016), 20 species have been identified, partially corresponding to those already reported in the literature (Donis-González et al. 2016; Magan 2006; Overy et al. 2003; Pietri et al. 2012; Sieber et al. 2007), with *P. crustosum* (Fig. 7.3), *P. glabrum* and *P. bialowiezense* as predominant species (57% of the isolates). Other species less frequently found, but often isolated from other types of nuts, were *P. expansum*, *P. palitans*, *P. chrysogenum* and *P. discolor* (Mujica and Vergara 1945; Frisvad and Samson 2004; Donis-González et al. 2016). For the first time the species *P. bialowiezense*, *P. brevicompactum*, *P. citrinum*, *P. commune*, *P. glandicola*, *P. manginii*, *P. pancosmium*, *P. polonicum*, *P. solitum*, *P. viridicatum*, *P. verrucosum*, *P. nordicum* and *P. palitans* have been reported as pathogens on chestnut. Seventy percent of the analysed strains were virulent on chestnut, including the indoor strains, according to results of Louw and Korsten (2014) on apples and pears. The most virulent species, according to data reported in literature, were *P. expansum* and *P. crustosum*, fungi that are also reported to be able to adapt to different environments (Louw and Korsten 2014; Scholtz and Korsten 2016).

Fig. 7.3 A chestnut showing blue mould, caused by *Penicillium crustosum*

The ability of the selected *Penicillium* species to produce patulin, ochratoxin A, penitrem A, roquefortine C, meleagrin, glandicolin A and B, mycophenolic acid, verrucosidin, andrastin A and penicillic acid *in vitro* was verified using high performance liquid chromatography coupled with a mass detector. The results obtained showed that different secondary metabolites were produced in CYA, especially at 10 and 14 days after inoculation. Furthermore, higher number and quantity of mycotoxins were found in the fungal mycelium compared to those present in the medium. It should be considered that different factors besides humidity, light and growth medium could influence the biosynthesis of secondary metabolites, which are often not activated under *in vitro* growth conditions (Brakhage 2013). Fifty-nine percent of the strains were able to produce at least one mycotoxin on chestnut, and, in total, 14 secondary metabolites were found on inoculated chestnuts. Furthermore, in a further experiment, the presence of penitrem A, roquefortine C, patulin, andrastin A, cyclopenin, cyclopenol and chaetoglobosin A was observed in 24 chestnut samples stored for 6 months, confirming the potential risk of chestnuts contamination. It should be emphasized that none of these molecules are currently regulated by European legislation on chestnuts.

Conclusion

A significant fraction of nut production is lost due to direct fungal contamination or production of regulated mycotoxins, such as aflatoxins. The main fungal genus involved in this loss of product is *Aspergillus*, which can often contaminate nuts during the supply chain, from harvest to storage. Beyond the mere economic damage, species of *Penicillium* can produce toxic compound currently not regulated in Europe, such as penitrem A, roquefortine C, patulin, andrastin A, cyclopenin, cyclopenol and chaetoglobosin A. It is therefore necessary to develop harvesting methods, storage conditions and decontamination and detoxification protocols able to reduce product loss and guarantee food safety for consumers. In particular, it is of paramount importance to develop storage conditions that avoid unstable temperatures or inefficient dehumidification, both of which cause a rapid increase in water activity and a consequent proliferation of filamentous fungi. Regarding chestnut production, increasing drying temperatures from 30 °C to 45 °C seems to be a significant step in the direction to reduce *Aspergillus* contamination and aflatoxin production. For hazelnuts, a possible improvement over existing protocols would be to favour machine-assisted rapid drying over sun drying, diminishing the fungal growth and therefore mycotoxin contamination. In hazelnuts dried at 45 °C, no aflatoxins were found after drying and storage, suggesting that this temperature of drying could be chosen as a compromise between quality and safety of nuts. Sometimes aflatoxins production cannot be entirely avoided, so it is useful to develop detoxification protocols able to reduce mycotoxins without diminishing food quality: infra-red roasting and utilization of cold plasma are both promising methods.

Acknowledgements This research was funded by Piedmont Region with the project "Food Digital Monitoring – FDM" (Framework Program MIUR – Piedmont Region, Action 3 "Smart Factory", Call "Smart Factory Technological Platform") and by CRT Foundation with the project "INNOCHEST – Innovative technologies to guarantee the quality and safety of Piedmontese chestnut".

References

Abdel-Gawad KM, Zohri AA (1993) Fungal flora and mycotoxins of six kinds of nut seeds for human consumption in Saudi Arabia. Mycopathologia 124:55–64

Abdel-Hafez AII, Sabah MS (1993) Mycoflora and mycotoxin of hazelnut (*Corylus avellana* L.) and walnut (*Juglans regia* L.) seeds in Egypt. Zentralbl Mikrobiol 148:137–147

Abrar M, Anjum FM, Butt MS, Pasha I, Randhawa MA, Saeed F, Waqas K (2012) Aflatoxins: biosynthesis, occurrence, toxicity, and remedies. Crit Rev Food Sci Nutr 53:862–874

Alasalvar C, Shahidi F (2008) Tree nuts: composition, phytochemicals, and health effects. CRC Press, Boca Raton

Atehnkeng J, Ojiambo PS, Cotty PJ, Bandyopadhyay R (2014) Field efficacy of a mixture of atoxigenic *Aspergillus flavus* Link: Fr vegetative compatibility groups in preventing aflatoxin contamination in maize (*Zea mays* L.). Biol Control 72:62–70

Attanasio G, Cinquanta L, Albanese D, Di Matteo M (2004) Effects of drying temperatures on physico-chemical properties of dried and rehydrated chestnuts (*Castanea sativa*). Food Chem 88:583–590

Aycicek H, Aksoy A, Saygi S (2005) Determination of aflatoxin levels in some dairy and food products which consumed in Ankara, Turkey. Food Control 16:263–266

Baltaci C, Ilyasoğlu H, Cavrar S (2012) Aflatoxin levels in raw and processed hazelnuts in Turkey. Food Addit Contam B 5:83–86

Barreto MC, Houbraken J, Samson RA, Frisvad JC, San-Romão MV (2011) Taxonomic studies of the *Penicillium glabrum* complex and the description of a new species *P. subericola*. Fungal Divers 49:23–33

Barros AIRNA, Nunes FM, Gonçalves B, Bennett RN, Silva AP (2011) Effect of cooking on total vitamin C contents and antioxidant activity of sweet chestnuts (*Castanea sativa* Mill.). Food Chem 128:165–172

Bayman P, Baker JL, Mahoney NE (2002) *Aspergillus* on tree nuts: incidence and associations. Mycopathologia 155:161–169

Bertuzzi T, Rastelli S, Pietri A (2015) *Aspergillus* and *Penicillium* toxins in chestnuts and derived produced in Italy. Food Control 50:876–880

Brakhage AA (2013) Regulation of fungal secondary metabolism. Nat Rev Microbiol 11:21–32

Campbell BC, Molyneux RJ, Schatzki TF (2003) Current research on reducing pre-and post-harvest aflatoxin contamination of US almond, pistachio, and walnut. J Toxicol Toxin Rev 22:225–266

CAST (2003) Mycotoxins: risks in plant, animal, and human systems. Task Force Report No. 139. Council for Agricultural Science and Technology, Ames

Codex Alimentarius (2010) CAC/RCP 59-2005. Code of practice for the prevention and reduction of aflatoxin in tree nuts. Joint FAO/WHO Food Standards Programme. Codex Alimentarius Commission, Rome

Di Stefano V, Pitonzo R, Cicero N, D'Oca MC (2014) Mycotoxin contamination of animal feedingstuff: detoxification by gamma-irradiation and reduction of aflatoxins and ochratoxin A concentrations. Food Addit Contam A 31:2034–2039

Donis-González IR, Guyer DE, Fulbright DW (2016) Quantification and identification of microorganisms found on shell and kernel of fresh edible chestnuts in Michigan. J Sci Food Agric 96:4514–4522

Dorner JW, Cole RJ (2001) Effect of application of nontoxigenic strains of *Aspergillus flavus* and *A. parasiticus* on subsequent aflatoxin contamination of peanuts in storage. J Stored Prod Res 38:329–339

Doster MA, Michailides TJ (1994) *Aspergillus* molds and aflatoxins in pistachio nuts in California. Phytopathology 84:583–590

Doster MA, Michailides TJ (1999) Relationship between shell discoloration of pistachio nuts and incidence of fungal decay and insect infestation. Plant Dis 83:259–264

European Food Safety Authority (EFSA) (2007) Opinion of the scientific panel on contaminants in the food chain on a request from the commission related to the potential increase of consumer health risk by a possible increase of the existing maximum levels for aflatoxins in almonds, hazelnuts and pistachios and derived products. EFSA J 446:1–127

FAOSTAT (2020). http://www.fao.org/faostat/en/#home. Accessed 2 Apr 2020

Fontana M, Somenzi M, Tesio A (2012) Cultivation, harvest and postharvest aspects that influence quality and organoleptic properties of hazelnut production and related final products. Acta Hortic 1052:311–314

Fridman G, Brooks AD, Balasubramanian M, Fridman A, Gutsol A, Vasilets VN, Ayan H, Friedman G (2007) Comparison of direct and indirect effects of non-thermal atmospheric-pressure plasma on bacteria. Plasma Process Polym 4:370–375

Frisvad JC, Samson RA (2004) Polyphasic taxonomy of *Penicillium* subgenus *Penicillium*. A guide to identification of food and air-borne terverticillate Penicillia and their mycotoxins. Stud Mycol 49:1–174

Golge O, Hepsag F, Kabak B (2016) Determination of aflatoxins in walnut sujuk and Turkish delight by HPLC-FLD method. Food Control 59:731–736

Gürses M (2006) Mycoflora and aflatoxin content of hazelnuts, walnuts, peanuts, almonds and roasted chickpeas (LEBLEBI) sold in Turkey. Int J Food Prop 9:395–399

Horn BW (2005) Colonization of wounded peanut seeds by soil fungi: selectivity for species from *Aspergillus* section *Flavi*. Mycologia 97:202–217

Houbraken J, Samson RA (2011) Phylogeny of *Penicillium* and the segregation of *Trichocomaceae* into three families. Stud Mycol 70:1–51

Houbraken J, Frisvad JC, Seifert KA, Overy DP, Tuthill DM, Valdez JG, Samson RA (2012) New penicillin-producing *Penicillium* species and an overview of section *Chrysogena*. Persoonia 29:78–100

Houbraken J, Wang L, Lee HB, Frisvad JC (2016) New sections in *Penicillium* containing novel species producing patulin, pyripyropens or other bioactive compounds. Persoonia 36:299–314

Jermini M, Conedera M, Sieber TN, Sassella A, Schärer H (2006) Influence of fruit treatments on perishability during cold storage of sweet chestnuts. J Sci Food Agric 86:877–885

Kabak B (2016) Aflatoxins in hazelnuts and dried figs: occurrence and exposure assessment. Food Chem 211:8–16

Kabak B, Dobson AD, Var I (2006) Strategies to prevent mycotoxin contamination of food and animal feed: a review. Crit Rev Food Sci Nutr 46:593–619

Kader AA (2013) Impact of nut postharvest handling, de-shelling, drying and storage on quality. In: Kader AA (ed) Improving the safety and quality of nuts. Woodhead Publishing, Cambridge, pp 22–34

Karlovsky P, Suman M, Berthiller F, De Meester J, Eisenbrand G, Perrin I, Dussort P (2016) Impact of food processing and detoxification treatments on mycotoxin contamination. Mycotoxin Res 32:179–205

Lahouar A, Marin S, Crespo-Sempere A, Saïd S, Sanchis V (2016) Effects of temperature, water activity and incubation time on fungal growth and aflatoxin B1 production by toxinogenic *Aspergillus flavus* isolates on sorghum seeds. Rev Argent Microbiol 48:78–85

Laroussi M (1996) Sterilization of contaminated matter with an atmospheric pressure plasma. IEEE Trans Plasma Sci 24:128–135

Liu X, Guan X, Xing F, Dai X, Liu Y (2017) Effect of water activity and temperature on the growth of *Aspergillus flavus*, the expression of aflatoxin biosynthetic genes and aflatoxin production in shelled peanuts. Food Control 82:325–332

López A, Piqué MT, Boatella J, Parcerisa J, Romero A, Ferrá A, García J (1997a) Influence of drying conditions on the hazelnut quality. I. Lipid oxidation. Dry Technol 15:965–977

López A, Piqué MT, Ferrán A, Romero A, Boatella J, García J (1997b) Influence of drying conditions on the hazelnut quality: II. Enzymatic activity. Dry Technol 15:979–988

López A, Piqué MT, Boatella J, Parcerisa J, Romero A, Ferrá A, García J (1997c) Influence of drying conditions on the hazelnut quality: III. Browning. Dry Technol 15:979–988

Louw JP, Korsten L (2014) Pathogenic *Penicillium* spp. on apples and pears. Plant Dis 98:590–598

Magan N (2006) Mycotoxin contamination of food in Europe: early detection and prevention strategies. Mycopathologia 162:245–253

Mao J, He B, Zhang L, Li P, Zhang Q, Ding X, Zhang W (2016) A structure identification and toxicity assessment of the degradation products of aflatoxin B1 in peanut oil under UV irradiation. Toxins 8:332

Marín S, Ramos AJ, Sanchis V (2012) Modelling *Aspergillus flavus* growth and aflatoxins production in pistachio nuts. Food Microbiol 32:378–388

Moreau M, Orange N, Feuilloley MGJ (2008) Non-thermal plasma technologies: new tools for bio-decontamination. Biotechnol Adv 26:610–617

Moreira R, Chenlo F, Chaguri L, Vázquez G (2011) Air drying and colour characteristics of chestnuts pre-submitted to osmotic dehydration with sodium chloride. Food Bioprod Process 89:109–115

Mujica F, Vergara C (1945) Flora fungosa Chilena. Indice preliminar de los huespedes de los hongos chilenos y sus referencias bibliograficas. Imprenta Stanley, Santiago

Nazzaro M, Barbarisi C, La Cara F, Volpe MG (2011) Chemical and biochemical characterisation of an IGP ecotype chestnut subjected to different treatments. Food Chem 128:930–936

Nielsen KF (2003) Mycotoxin production by indoor molds. Fungal Genet Biol 39:103–117

Overy DP, Seifert KA, Savard ME, Frisvad JC (2003) Spoilage fungi and their mycotoxins in commercially marketed chestnuts. Int J Food Microbiol 88:69–77

Ozay G, Seyhan F, Pembeci C, Saklar S, Yilmaz A (2008) Factors influencing fungal and aflatoxin levels in Turkish hazelnuts (*Corylus avellana* L.) during growth, harvest, drying and storage: a 3-year study. Food Addit Contam A 25:209–218

Özdemir M, Devres O (1999) Turkish hazelnuts: properties and effect of microbiological and chemical changes on quality. Food Rev Int 15:309–333

Özilgen M, Özdemir M (2001) A review on grain and nut deterioration and design of the dryers for safe storage with special reference to Turkish hazelnuts. Crit Rev Food Sci Nutr 41:95–132

Pietri A, Rastelli S, Mulazzi A, Bertuzzi T (2012) Aflatoxins and ochratoxin A in dried chestnuts and chestnut flour produced in Italy. Food Control 25:601–606

Pildain MB, Vaamonde G, Cabral D (2004) Analysis of population structure of *Aspergillus flavus* from peanut based on vegetative compatibility, geographic origin, mycotoxin and sclerotia production. Int J Food Microbiol 93:31e40

Pitt JI, Hocking AD (2009) Fungi and food spoilage, 3rd edn. Springer, Dordrecht

Prelle A, Spadaro D, Garibaldi A, Gullino ML (2012) Aflatoxin monitoring in Italian hazelnut products by LC-MS. Food Addit Contam B 5:279–285

Prencipe S, Siciliano I, Contessa C, Botta R, Garibaldi A, Gullino ML, Spadaro D (2018a) Characterization of *Aspergillus* section *Flavi* isolated from fresh chestnuts and along the chestnut flour process. Food Microbiol 69:159–169

Prencipe S, Siciliano I, Gatti I, Gullino ML, Garibaldi A, Spadaro D (2018b) Chestnut drying is critical in determining *Aspergillus flavus* growth and aflatoxin contamination. Toxins 10:530

Prencipe S, Siciliano I, Gatti C, Garibaldi A, Gullino ML, Botta R, Spadaro D (2018c) Several species of *Penicillium* isolated from chestnut flour processing are pathogenic on fresh chestnuts and produce mycotoxins. Food Microbiol 76:396–404
Probst C, Cotty PJ (2012) Relationships between *in vivo* and *in vitro* aflatoxin production: reliable prediction of fungal ability to contaminate maize with aflatoxins. Fungal Biol 116:503e510
Rachaputi NR, Wright GC, Kroschi S (2002) Management practices to minimise pre-harvest aflatoxin contamination in Australian groundnuts. Aust J Exp Agric 42:595–605
Rodrigues P, Venâncio A, Kozakiewicz S, Lima N (2009) A polyphasic approach to the identification of aflatoxigenic and non-aflatoxigenic strains of *Aspergillus* section *Flavi* isolated from Portuguese almonds. Int J Food Microbiol 129:187–193
Rodrigues P, Venâncio A, Lima N (2012) Mycobiota and mycotoxins of almonds and chestnuts with special reference to aflatoxins. Food Res Int 48:76–90
Rodrigues P, Venâncio A, Lima N (2013) Incidence and diversity of the fungal genera *Aspergillus* and *Penicillium* in Portuguese almonds and chestnuts. Eur J Plant Pathol 137:197–209
Ros E (2010) Health benefits of nuts consumption. Nutrients 2:652–682
Samson RA, Visagie CM, Houbraken J, Hong S-B, Hubka V, Klaassen CHW, Perrone G, Seifert KA, Susca A, Tanney JB, Varga J, Kocsubé S, Szigeti G, Yaguchi TY, Frisvad JC (2014) Phylogeny, identification and nomenclature of the genus *Aspergillus*. Stud Mycol 78:141–173
Schmidt-Heydt M, Rüfer CE, Abdel-Hadi A, Magan N, Geisen R (2010) The production of aflatoxin B1 or G1 by *Aspergillus parasiticus* at various combinations of temperature and water activity is related to the ratio of *aflS* to *aflR* expression. Mycotoxin Res 26:241–246
Scholtz I, Korsten L (2016) Profile of *Penicillium* species in the pear supply chain. Plant Pathol 65:1126–1132
Scott JA, Wong B, Summerbell RC, Untereiner WA (2008) A survey of *Penicillium brevicompactum* and *P. bialowiezense* from indoor environments, with commentary on the taxonomy of the *P. brevicompactum* group. Can J Bot 86:732–741
Selcuk M, Oksuz L, Basaran P (2008) Decontamination of grains and legumes infected with *Aspergillus* spp. and *Penicillium* spp. by cold plasma treatment. Bioresour Technol 99:5104–5109
Shantha T (1999) Fungal degradation of aflatoxin B1. Nat Toxins 7:175–178
Siciliano I, Spadaro D, Prelle A, Vallauri D, Cavallero MC, Garibaldi A, Gullino ML (2016) Use of cold atmospheric plasma to detoxify hazelnuts from aflatoxins. Toxins 8:125
Siciliano I, Dal Bello B, Zeppa G, Spadaro D, Gullino ML (2017) Static hot air and infrared rays roasting are efficient methods for aflatoxin decontamination on hazelnuts. Toxins 9:72
Sieber TN, Jermini M, Conedera M (2007) Effects of the harvest method on the infestation of chestnuts (*Castanea sativa*) by insects and moulds. J Phytopathol 155:497–504
Şimşek O, Arici M, Demir C (2002) Mycoflora of hazelnut (*Corylus avellana* L.) and aflatoxin content in hazelnut kernels artificially infected with *Aspergillus parasiticus*. Food Nahrung 46:194–196
Spadaro D, Garibaldi A (2017) Containment of mycotoxins in the food chain by using decontamination and detoxification techniques. In: Gullino ML, Stack J, Fletcher J, Mumford J (eds) Practical tools for plant and food biosecurity. Springer, Dordrecht, pp 163–177
Thomson SV, Mehdy MC (1978) Occurrence of *Aspergillus flavus* in pistachio nuts prior to harvest. Phytopathology 68:1112–1114
Trenk HL, Harman PA (1970) Effects of moisture content and temperature on aflatoxin production in corn. Appl Microbiol 19:781–784
Turan A (2018) Effect of drying methods on fatty acid profile and oil oxidation of hazelnut oil during storage. Eur Food Res Technol 244:2181–2190
Turan A, Karaosmanoğlu H (2019) Effect of drying methods on long term storage of hazelnut. Food Sci Technol 39:406412
Vaamonde G, Patriarca A, Pinto VF, Comeria R, Degrossi C (2003) Variability of aflatoxin and cyclopiazonic acid production by *Aspergillus* section *Flavi* from different substrates in Argentina. Int J Food Microbiol 88:79e84

Valente S, Meloni GR, Prencipe S, Spigolon N, Somenzi M, Fontana M, Gullino ML, Spadaro D (2020) Effect of drying temperature and exposure times on *Aspergillus flavus* growth and aflatoxin production on artificially inoculated hazelnuts. J Food Prot. https://doi.org/10.4315/JFP-20-061

Visagie CM, Hirooka Y, Tanney JB, Whitfield E, Mwange K, Meijer M, Amend AS, Seifert KA, Samson RA (2014a) *Aspergillus*, *Penicillium* and *Talaromyces* isolated from in house dust samples collected around the world. Stud Mycol 78:63–139

Visagie CM, Houbraken J, Frisvad JC, Hong SB, Klaassen CHW, Perrone G, Seifert KA, Varga J, Yaguchi T, Samson RA (2014b) Identification and nomenclature of the genus *Penicillium*. Stud Mycol 78:343–371

Washington WS, Allen AD, Dooley LB (1997) Preliminary studies on *Phomopsis castanea* and other organisms associated with healthy and rotted chestnut fruit in storage. Australas Plant Pathol 26:37–43

Wells JM, Payne JA (1975) Toxigenic *Aspergillus* and *Penicillium* isolates from weevil-damaged chestnuts. Appl Microbiol 30:536–540

Zhu F (2011) Effect of processing on quality attributes of chestnut. Food Biotechnol 9:1429–1443

Chapter 8
The Wheat Microbiome in Relation to Mycotoxin Occurrence in Stored Grain: An Overview

Manoj Kumar Solanki, Samir Droby, and Edward Sionov

Abstract Wheat is the primary source of carbohydrates and considered as a widely consumed food worldwide. Due to its biologic composition and structure, wheat grains are one of the more highly food sources that can be stored and used for a long time. Preservation techniques have been used to extend the shelf life of wheat and wheat-based products, primarily to reduce the growth and proliferation of mycotoxigenic pathogens and the development of fungal spoilage. Several preservation techniques are well established to preserve the wheat for a long duration, but numerous outbreaks regarding spoiled wheat and wheat-based products are reported each year. Wheat grains are colonized by complex microbial communities that play different roles in grain quality and susceptibility to disease. Some of the bacteria and fungi that are associated with seeds can be harmful to human health and cause plant diseases, while others can have beneficial effects on the host and actually improve nutrient uptake and increase tolerance to biotic and abiotic stresses through multiple mechanisms. A better knowledge of the composition and dynamics of wheat-grain-associated microbiota is needed to identify novel beneficial microorganisms that may improve crop health and suppress the growth of potential pathogens in a sustainable manner.

Keywords Stored wheat grain microbiome · Mycotoxigenic fungi · Mycotoxins

M. K. Solanki · S. Droby (✉) · E. Sionov (✉)
Agricultural Research Organization, Institute of Postharvest and Food Sciences, The Volcani Center, Rishon LeZion, Israel
e-mail: samird@volcani.agri.gov.il; edwardsio@volcani.agri.gov.il

D. Spadaro et al. (eds.), *Postharvest Pathology*, Plant Pathology in the 21st Century 11, https://doi.org/10.1007/978-3-030-56530-5_8

Introduction

Wheat is one of the essential cultivated grain crops and is an important source of calories and plant-derived protein in human food. The availability of high-quality wheat grains is essential for maintaining the stability of the world's food supply and for global food security. Wheat grains are colonized by complex microbial communities that play different roles in grain quality and susceptibility to disease (Links et al. 2014). Some of the bacteria and fungi that are associated with seeds can be harmful to human health and cause plant diseases, while others can have beneficial effects on the host and actually improve nutrient uptake and increase tolerance to biotic and abiotic stresses through multiple mechanisms (Brader et al. 2017). When grains are colonized by moulds there is a significant risk of contamination with mycotoxins, which are toxic chemical products formed as secondary metabolites by these fungi. Many species of *Fusarium*, *Aspergillus*, *Penicillium*, and *Alternaria* are not only recognized as plant pathogens but are also sources of important mycotoxins of concern in relation to animal and human health (Magan et al. 2010; Placinta et al. 1999). Poor postharvest management can lead to rapid deterioration in quality, with severe decreases in germinability and nutritional value of stored grain, possibly accompanied by undesirable fungal contamination and, consequently, toxin production (Magan et al. 2003). Interactions between environmental stress factors, such as high moisture content (water activity) and temperature stress, may have an influence on growth, expression of biosynthetic regulatory genes and mycotoxin production by mycotoxigenic fungal species (Bernáldez et al. 2017; Medina et al. 2015). These fungi may cause significant wheat grain yield losses in the field and storage facilities, especially due to their ability to produce mycotoxins, making crops unsafe for consumption (Chen et al. 2019; Ferrigo et al. 2016; Lee and Ryu 2017; McCormick et al. 2011; Streit et al. 2013). In the United States alone, the crop losses from mycotoxin contamination averaged $932 million every year (CAST 2003). Timely assessment of these contaminants and identification of the main toxicogenic fungal species are important, not only for assessing food quality but also for the development of control strategies for ensuring food safety (Sadhasivam et al. 2017; Suanthie et al. 2009). In addition, the plant protective ability of some bacterial epiphytes and endophytes against fungal pathogens has been reported in several crops (Gdanetz and Trail 2017; Links et al. 2014; Mousa et al. 2016). A better knowledge of the composition and dynamics of wheat-grain-associated microbiota is needed to identify novel beneficial microorganisms that may improve crop health and suppress the growth of potential pathogens in a sustainable manner.

Wheat Grain Microbial Communities

Bacterial Microbiota

During the growing season, wheat and other cereal grains are exposed to different sources of microbial contamination, including soil, water, insects, birds, cropping equipment, humans and animals (Sabillón and Bianchini 2016). The bacterial microflora found in the wheat grain is diverse and mainly belong to the families *Micrococcaceae*, *Pseudomonadaceae*, *Enterobacteriaceae*, *Lactobacillaceae*, *Bacillaceae*, and, within these families, pathogenic species such as *Bacillus cereus*, *Clostridium botulinum*, *Clostridium perfringens*, *Escherichia coli*, *Salmonella*, and *Staphylococcus aureus* are found (Berghofer et al. 2003; Eglezos 2010; Sabillón and Bianchini 2016). In addition, microbial contamination also occurs during harvesting and subsequent handling and storage. In a recent study, amplicon-based high-throughput sequencing was used to characterize the wheat grain microbiome at different time points over a storage period of 6 months (Solanki et al. 2019). In this study, bacterial communities of stored grains mainly consisted of genera of *Bacillus*, followed by *Erwinia*, *Pseudomonas*, unidentified genera that belong to the family *Oxalobacteraceae*, *Enterobacteriaceae*, *Streptomyces*, *Curtobacterium* and *Paenibacillus* (Fig. 8.1). The majority of the detected bacteria taxa, with the exception of *Erwinia*, are known to have a beneficial impact on plants. For instance, *Bacillus* species are known to have a promotive growth effect on wheat plants and are generally isolated from both grains and the rhizosphere (Cherif-Silini et al. 2016; Pan et al. 2015; Tao et al. 2014). *Paenibacillus* is recognized as a predominantly endophytic bacterium in wheat plants and seeds, also with promotive growth effects (Díaz Herrera et al. 2016; Hao and Chen 2017). Similarly, *Pseudomonas* contains several species that promote plant growth by suppressing pathogenic microorganisms and synthesizing growth-stimulating plant hormones (Preston 2004). Notably, however, other species of *Pseudomonas*, as well as Bacillus, may be phytopathogenic. Moreover, Solanki et al. (2019) also found a correlation between the examined cultured bacterial community of stored wheat grains and the results obtained by high-throughput sequencing data. *Bacillus*, *Pseudomonas*, and *Pantoea* were found to be the most abundant genera obtained in the cultured bacterial taxa isolated from stored wheat grains throughout the study, and the same was true in the high-throughput sequencing analysis.

Fungal Microbiota

The type and amount of fungal microorganisms present in wheat are affected by several factors, such as weather conditions prevailing during grain ripening and harvesting, moisture content, and insect, bird, and rodent activity during production and storage (Doyle and Buchanan 2013). Traditionally, fungal contaminants in crops have been divided into two distinct classes: 'field fungi' (or plant pathogens),

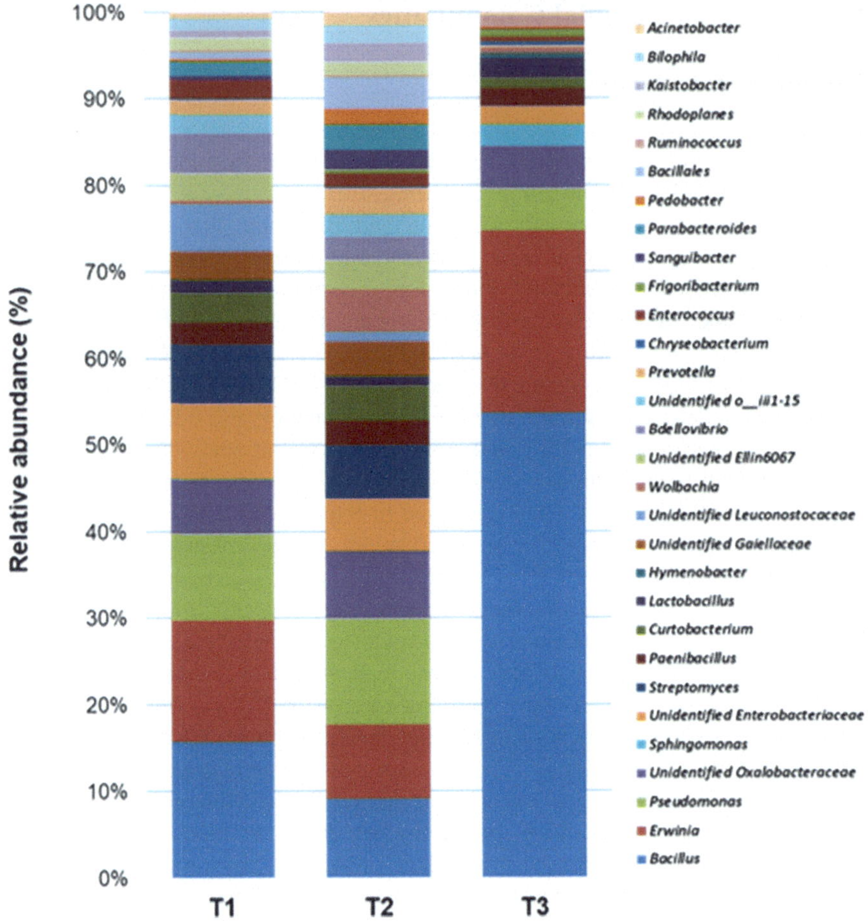

Fig. 8.1 Relative abundance of bacterial (⩾0.2%) microbiota at the genus level detected using high-throughput sequencing technology across the wheat grain samples collected at three storage time-points (T1, T2, T3). (Solanki et al. 2019)

which invade and produce mycotoxins before harvest, and 'storage' (or saprophytic) fungi, which become a problem after harvest. However, the original source of both of these classes of fungi is in the field (Miller 1995). Moreover, some fungi might belong to both classes and colonize grains before and after harvest. *Aspergillus flavus*, for example, is associated with *Aspergillus* infections of grain crops in the field; it also contaminates stored grains when prevailing abiotic factors (such as temperature and water activity) are favorable (Campbell and White 1995; Smith 1997). The fungal mycobiota found in the wheat grains includes yeasts and molds that mostly belong to *Alternaria*, *Cladosporium*, *Fusarium*, *Helminthosporium*, *Aspergillus*, *Penicillium*, and *Eurotium* (Doyle and Buchanan 2013; Laca et al. 2006). Yuan et al. (2018) reported that *Alternaria*, *Filobasidium*, *Fusarium*, and *Wallemia* are the most abundant fungal genera that were isolated from freshly

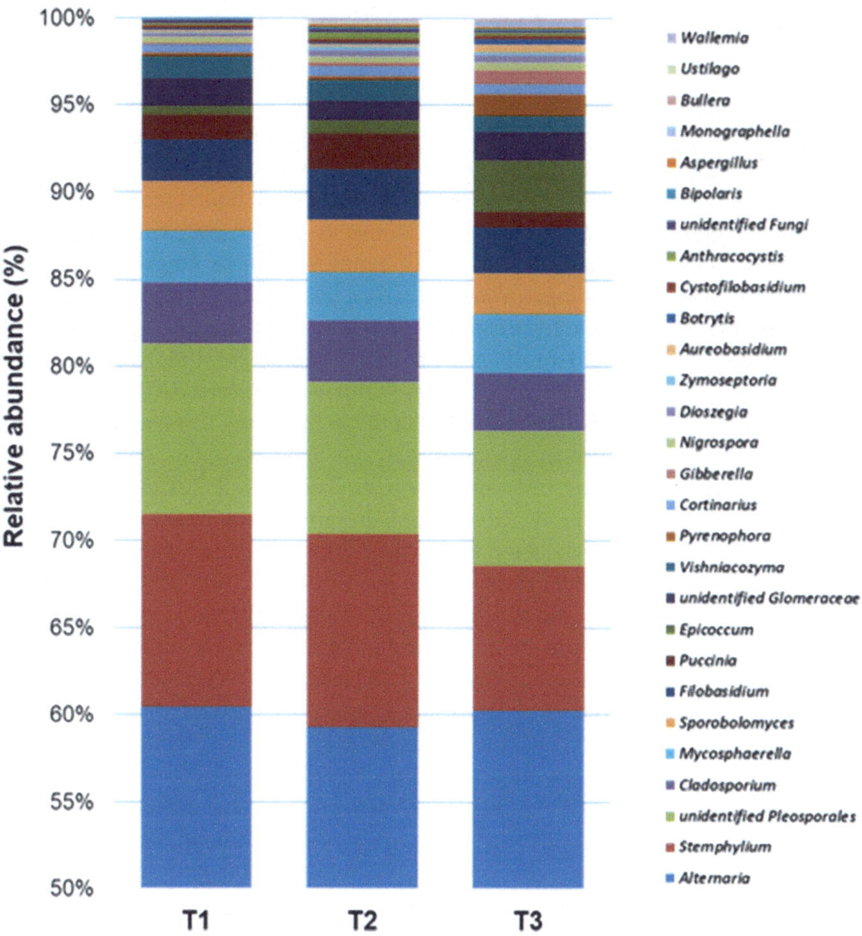

Fig. 8.2 Relative abundance of fungal (≥0.1%) genera detected using high-throughput sequencing technology across the wheat grain samples collected at three storage time-points (T1, T2, T3). (Solanki et al. 2019)

harvested wheat. The analysis of stored wheat grain mycobiota, using high-throughput amplicon sequencing of the fungal internal transcribed spacer (ITS) region (Fig. 8.2), revealed that *Alternaria*, *Stemphylium*, unidentified *Pleosporales*, *Cladosporium*, *Mycosphaerella*, *Sporobolomyces*, *Filobasidium*, and *Puccinia* were the most abundant fungal genera, which accounted for more than 90% of the total fungal community and were equally present in all three of the sampled points (Solanki et al. 2019). *Cryptococcus* and *Rhodotorula* species predominated among the isolated yeast cultures from stored grains based on morphological features and sequencing analysis, while *Alternaria*, *Aspergillus*, *Fusarium*, *Massarina*, *Penicillium*, and *Phoma* were the most abundant species cultured from stored wheat grains (Solanki et al. 2019).

Mycotoxigenic Fungi and Mycotoxins in Wheat Grains

Mycotoxins are low molecular weight natural products produced as secondary metabolites by toxigenic filamentous fungi that contaminate food and the food chain, and represent a risk to human and animal health. The major mycotoxins that occur in wheat-based retail products, especially whole grain flours, are produced by *Fusarium*, *Aspergillus* and/or *Penicillium* (Cabañas et al. 2008; Gashgari et al. 2010). When present in foods at sufficiently high levels, these fungal metabolites can have a toxic effect that ranges from acute (for example liver or kidney deterioration) to chronic (for example liver cancer). Such levels can also be mutagenic and/or teratogenic, which could result in symptoms ranging from skin irritation to immunosuppression, birth defects, neurotoxicity, and death (ICMSF 1996). Mycotoxins can enter the food chain either in the field, during storage, or at a later point. The mycotoxin problem is exacerbated whenever shipping, handling, and storage practices are conducive to mold growth. This contributes to mycotoxins being commonly found in foods. Mycotoxins are now recognized to be the most important chronic dietary risk factor, higher than synthetic contaminants, plant toxins, food additives or pesticide residues (Kuiper-Goodman 1998). In addition, the presence of mycotoxins in grains is also an economic concern. A quarter of the world's crops are estimated to be contaminated to some extent with mycotoxins (Fink-Gremmels 1999). The most frequently detected mycotoxins in wheat grain are deoxynivalenol (DON), fumonisins and zearalenone, produced by *Fusarium* species during wheat flowering (*F. graminearum*, *F. culmorum*, *F. verticillioides* and *F. proliferatum*), and aflatoxins and ochratoxin A (OTA), produced by *Aspergillus* (such as *A. ochraceus*) and *Penicillium* (*P. verrucosum*) species, respectively, during wheat storage (Magan et al. 2010; Priyanka et al. 2015). Among the mycotoxins produced by *Fusarium* spp., DON is the predominant and most economically important mycotoxin in small grain production, such as wheat (Wegulo 2012). Due to their wide range of physical and chemical properties, mycotoxins are stable chemical compounds that cannot be destroyed during most food processing operations. Regulatory agencies such as the European Commission have established maximum limits for these toxins to minimize the exposure and toxic effects in humans and animals (EC 2006a, b, c).

Shifts in Mycotoxin Profiles Due to Changes in Microbiome Composition of Stored Grains

In addition to fungal pathogens that cause seed deterioration during storage, insects can infest and spoil grain that is handled or stored improperly. Therefore, fumigation treatment is often required to prevent insect infestations in stored grains. Phosphine gas is by far the most common fumigation agent used for stored grain insect control and is considered one of the most effective insect control measures when properly applied (Phillips et al. 2012). A recent study of Solanki et al. (2019) identified significant changes in the composition of the microbial community when phosphine fumigation was applied to stored wheat grains. This study demonstrated that phosphine fumigation significantly affected the microbial community associated with wheat grains. Since phosphine is toxic to aerobically respiring organisms, it also has effects on the survival and growth of some aerobic bacteria and fungi (Hocking and Banks 1991; Castro et al. 2001). Bacterial population exhibited a drastic reduction in their level of diversity and number of observed species after fumigation. Highly abundant bacteria, including *Streptomyces*, *Bdellovibrio*, *Prevotella*, *Bilophila*, *Parabacteroides* and *Leuconostocaceae*, that were identified by high-throughput sequencing analysis of stored wheat grain samples prior fumigation, disappeared entirely after the phosphine treatment (Solanki et al. 2019). This change corresponded with a significant increase in *Bacillus* from an average of 2.2% prior fumigation (first two-time points - T1 and T2) to more than 50% after phosphine treatment (last time point - T3) (Fig. 8.1). A similar trend was reported by Yuan et al. (2018), who reported that the abundance of *Bacillus* spp. in wheat grains increased from 2% to 36% after 9 months of storage and up to 48% after 12 months of storage at a different position within a storage silo. Interestingly, in that study, the authors indicated that stored wheat grain was treated with ozone to kill insects. Small quantities of stable but harmless breakdown products of phosphine, which become incorporated into normal cellular metabolism as phosphates and phosphites, may remain in fumigated materials (Nath et al. 2011). *Bacillus* spp. are capable of utilizing phosphite and belong to the group of phosphate solubilizing bacteria, which play an important role in plant growth promotion (Foster et al. 1978; Igual et al. 2001; Kang et al. 2014). *Bacillus* may take advantage of these unique abilities after phosphine fumigation to enhance their survival and proliferation.

Only a few studies have described the influence of phosphine on the survival and growth of molds in stored grains. Phosphine was reported to have little effect on populations of fungi that were unable to grow in stored grains under conditions of limiting water activity (a_w 0.8–0.86) (Sinha et al. 1967; Raghunathan et al. 1969; Hocking and Banks 1991). Natarajan and Bagyaraj (1984) noted some reduction in fungal growth on legumes when exposed to very high phosphine levels (100 g/m^3), particularly at limiting a_w conditions (0.8). It seems that the effect of phosphine on fungal growth, however, may vary between fungal species and the type of stored grain or seed. Phosphine was shown to inhibit the growth of *A. flavus* and/or *A. parasiticus* and aflatoxin production on peanuts in both laboratory and warehouse experiments (Castro et al. 1992, 1995, 1996). A reduction of *A. flavus* growth and aflatoxin production was also observed in response to phosphine treatment of maize kernels stored at different moisture levels (Castro et al. 2001). In the same study, however, the authors stated that *Penicillium* species and *F. verticillioides*, which are commonly found in freshly harvested grains, were tolerant to phosphine (Castro et al. 2001). Solanki et al. (2019) reported that unlike the bacterial population, the fungal abundance and diversity in stored grain samples were not affected following phosphine treatment (Fig. 8.2). Moreover, a greater number of *Fusarium* species, such as *F. culmorum* and *F. proliferatum*, were detected in grain samples after

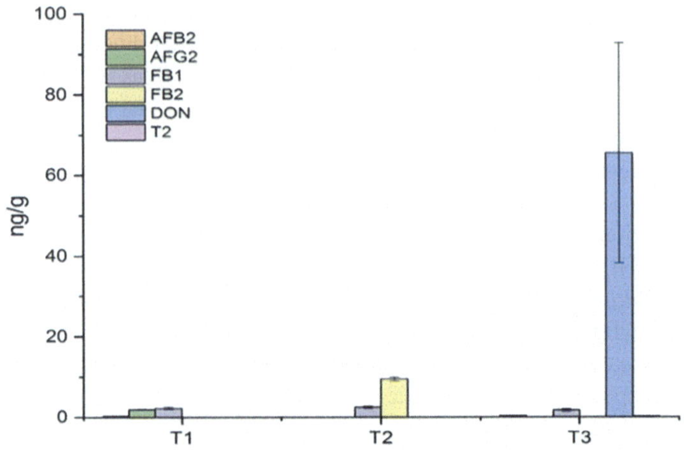

Fig. 8.3 Mycotoxin contamination in stored wheat grain samples at three storage time-points (T1, T2, T3). Error bars represent the standard error of the mean (SEM) across three independent replicates

fumigation, probably due to the changes that occurred in microbial community composition following phosphine treatment. Notably, the presence of mycotoxigenic *Fusarium* isolates in the stored wheat grains after fumigation (T3 time point) was strongly associated with the occurrence of DON in grain samples that underwent phosphine treatment, suggesting that phosphine induced a shift in the microbial composition towards more toxigenic strains (Fig. 8.3).

Conclusion

The development of next-generation sequencing technologies has provided researchers with a deeper understanding of the microbiome on plant material and with an opportunity to reveal large numbers of microbes, generating communities that can affect disease outcomes. Since many of the microorganisms identified by high-throughput sequencing technology are not culturable, it would be very difficult to detect alterations in microbial community diversity and abundance after different management practices using only classical culturing and molecular identification techniques. Several studies have reported that high-throughput molecular approaches were able to identify changes in the microbiome that were not detectable using classical methodologies (Sylla et al. 2013; Schmidt et al. 2014). More research is required to better understand the interactions that occur among the different members of the microbial community of stored wheat grains under different management strategies, which could assist in the prediction of fungal disease incidence and mycotoxin production, as well as in the development of novel approaches for controlling mycotoxin contamination in grains and managing crop diseases in general.

References

Berghofer LK, Hocking AD, Miskelly D, Jansson E (2003) Microbiology of wheat and flour milling in Australia. Int J Food Microbiol 85:137–149. https://doi.org/10.1016/S0168-1605(02)00507-X

Bernáldez V, Córdoba JJ, Magan N, Peromingo B, Rodríguez A (2017) The influence of eco-physiological factors on growth, aflR gene expression and aflatoxin B1 production by a type strain of *Aspergillus flavus*. LWT Food Sci Technol 83:283–291. https://doi.org/10.1016/j.lwt.2017.05.030

Brader G, Compant S, Vescio K, Mitter B, Trognitz F, Ma LJ, Sessitsch A (2017) Ecology and genomic insights into plant-pathogenic and plant-nonpathogenic endophytes. Annu Rev Phytopathol 55:61–83. https://doi.org/10.1146/annurev-phyto-080516-035641

Cabañas R, Bragulat MR, Abarca ML, Castellá G, Cabañes FJ (2008) Occurrence of *Penicillium verrucosum* in retail wheat flours from the Spanish market. Food Microbiol 25:642–647. https://doi.org/10.1016/j.fm.2008.04.003

Campbell KW, White DG (1995) Evaluation of corn genotypes for resistance to Aspergillus ear rot, kernel infection, and aflatoxin production. Plant Dis 79:1039–1045. https://doi.org/10.1094/PD-79-1039

CAST (2003) Mycotoxins: risks in plant, animal and human systems, Task force report, ISSN 0194-4088. Council for Agricultural Science and Technology, Ames

Castro MF, Pacheco IA, Taniwaki MH (1992) Effects of phosphine on aflatoxin production in peanuts stored with a high content. In: Methyl bromide technical options workshop, Washington, USA

Castro MF, Pacheco IA, Soares LM, Furlani RP (1995) Phosphine fumigation of stored peanuts for *Aspergillus flavus* Link/*Aspergillus parasiticus* Speare and aflatoxins control. Rev Microbiol 26:296–301

Castro MF, Pacheco IA, Soares LM, Furlani RP, De Paula DC, Bolonhezi S (1996) Warehouse control of *Aspergillus flavus* Link and *A. parasiticus* Speare on peanuts (*Arachis hypogaea*) by phosphine fumigation and its effect on aflatoxin production. J Food Prot 59:407–411. https://doi.org/10.4315/0362-028X-59.4.407

Castro MF, De Leitao MF, Do Vale JO, Bragnolo N, Anichiareo ES, Mills KA (2001) Effects of phosphine in the development of *Aspergillus flavus* aflatoxin production in maize grains stored at different moisture contents. International Working Conference on Controlled Atmosphere and Fumigation in Stored Products 2000 29:179–191

Chen Y, Kistler HC, Ma Z (2019) *Fusarium graminearum* Trichothecene mycotoxins: biosynthesis, regulation, and management. Annu Rev Phytopathol 57:15–39. https://doi.org/10.1146/annurev-phyto-082718-100318

Cherif-Silini H, Silini A, Yahiaoui B, Ouzari I, Boudabous A (2016) Phylogenetic and plant-growth-promoting characteristics of Bacillus isolated from the wheat rhizosphere. Ann Microbiol 66:1087–1097. https://doi.org/10.1007/s13213-016-1194-6

Díaz Herrera S, Grossi C, Zawoznik M, Groppa MD (2016) Wheat seeds harbour bacterial endophytes with potential as plant growth promoters and biocontrol agents of *Fusarium graminearum*. Microbiol Res 186:37–43. https://doi.org/10.1016/J.MICRES.2016.03.002

Doyle MP, Buchanan RL (2013) Food microbiology: fundamentals and frontiers. American Society for Microbiology Press, Washington, DC

EC (2006a) Commission recommendation (EC) 2006/576. Off J Eur Union L229:7

EC (2006b) Commission regulation (EC) 1881/2006. Off J Eur Union L364:5e24

EC (2006c) Commission Regulation (EC) 401/2006. Off J Eur Union, L70, 12

Eglezos S (2010) Microbiological quality of wheat grain and flour from two mills in Queensland, Australia. J Food Prot 73:1533–1536. https://doi.org/10.4315/0362-028x-73.8.1533

Ferrigo D, Raiola A, Causin R (2016) Fusarium toxins in cereals: occurrence, legislation, factors promoting the appearance and their management. Molecules 21:627. https://doi.org/10.3390/molecules21050627

Fink-Gremmels J (1999) Mycotoxins: their implications for human and animal health. Vet Q 21:115–120. https://doi.org/10.1080/01652176.1999.9695005

Foster TL, Winans L, Helms SJ, Helms SJ (1978) Anaerobic utilization of phosphite and hypophosphite by *Bacillus* sp. Appl Environ Microbiol 35:937–944

Gashgari RM, Shebany YM, Gherbawy YA (2010) Molecular characterization of mycobiota and aflatoxin contamination of retail wheat flours from Jeddah markets. Foodborne Pathog Dis 7:1047–1054. https://doi.org/10.1089/fpd.2009.0506

Gdanetz K, Trail F (2017) The wheat microbiome under four management strategies, and potential for endophytes in disease protection. Phytobiomes 1:158–168. https://doi.org/10.1094/PBIOMES-05-17-0023-R

Hao T, Chen S (2017) Colonization of wheat, maize and cucumber by *Paenibacillus polymyxa* WLY78. PLoS One 12:e0169980. https://doi.org/10.1371/journal.pone.0169980

Hocking AD, Banks HJ (1991) Effects of phosphine fumigation on survival and growth of storage fungi in wheat. J Stored Prod Res 27:115–120. https://doi.org/10.1016/0022-474X(91)90021-4

ICMSF (1996) Toxigenic fungi: Aspergillus. In: Roberts TA, Baird-Parker AC, Tompkin RB (eds) Microorganisms in foods 5: microbiological specifications of food pathogens. International Commission on Microbiological Specifications for Foods. Blackie Academic and Professional. An Imprint of Chapman & Hall, London, pp 347–381

Igual J, Valverde A, Cervantes E, Velazquez E (2001) Phosphate-solubilizing bacteria as inoculants 979 for agriculture: use updated molecular techniques in their study. Agronomie 21:561–568

Kang SM, Radhakrishnan R, You YH, Joo GJ, Lee IJ, Lee KE, Kim JH (2014) Phosphate solubilizing *Bacillus megaterium* mj1212 regulates endogenous plant carbohydrates and amino acids contents to promote mustard plant growth. Indian J Microbiol 54:427–433. https://doi.org/10.1007/s12088-014-0476-6

Kuiper-Goodman T (1998) Food safety: mycotoxins and phycotoxins in perspective. In: Miraglia M, van Edmond H, Brera C, Gilbert J (eds) Mycotoxins and phycotoxins – developments in chemistry, toxicology and food safety. Alaken Inc., Fort Collins, Colo, pp 25–48

Laca A, Mousia Z, Díaz M, Webb C, Pandiella SS (2006) Distribution of microbial contamination within cereal grains. J Food Eng 72:332–338. https://doi.org/10.1016/j.jfoodeng.2004.12.012

Lee HJ, Ryu D (2017) Worldwide occurrence of mycotoxins in cereals and cereal-derived food products: public health perspectives of their cooccurrence. J Agric Food Chem 65:7034–7051. https://doi.org/10.1021/acs.jafc.6b04847

Links MG, Demeke T, Gräfenhan T, Hill JE, Hemmingsen SM, Dumonceaux TJ (2014) Simultaneous profiling of seed-associated bacteria and fungi reveals antagonistic interactions between microorganisms within a shared epiphytic microbiome on Triticum and Brassica seeds. New Phytol 202:542–553. https://doi.org/10.1111/nph.12693

Magan N, Hope R, Cairns V, Aldred D (2003) Post-harvest fungal ecology: impact of fungal growth and mycotoxin accumulation in stored grain. Eur J Plant Pathol 109:723–730. https://doi.org/10.1023/A:1026082425177

Magan N, Aldred D, Mylona K, Lambert RJW (2010) Limiting mycotoxins in stored wheat. Food Addit Contam Part A 27:644–650. https://doi.org/10.1080/19440040903514523

McCormick SP, Stanley AM, Stover NA, Alexander NJ (2011) Trichothecenes: from simple to complex mycotoxins. Toxins (Basel) 3:802–814. https://doi.org/10.3390/toxins3070802

Medina A, Schmidt-Heydt M, Rodríguez A, Parra R, Geisen R, Magan N (2015) Impacts of environmental stress on growth, secondary metabolite biosynthetic gene clusters and metabolite production of xerotolerant/xerophilic fungi. Curr Genet 61:325–334. https://doi.org/10.1007/s00294-014-0455-9

Miller JD (1995) Fungi and mycotoxins in grain: implications for stored product research. J Stored Prod Res 31:16

Mousa WK, Shearer C, Limay-Rios V, Ettinger CL, Eisen JA, Raizada MN (2016) Root-hair endophyte stacking in finger millet creates a physicochemical barrier to trap the fungal pathogen *Fusarium graminearum*. Nat Microbiol 1:16167. https://doi.org/10.1038/nmicrobiol.2016.167

Natarajan T, Bagyaraj DJ (1984) Fumigation effect on micro flora and viability of black gram and field bean seeds. Pesticides 18:40–42

Nath NS, Bhattacharya I, Tuck AG, Schlipalius DI, Ebert PR (2011) Mechanisms of phosphine toxicity. J Toxicol 2011:494168. https://doi.org/10.1155/2011/494168

Pan D, Mionetto A, Tiscornia S, Bettucci L (2015) Endophytic bacteria from wheat grain as biocontrol agents of *Fusarium graminearum* and deoxynivalenol production in wheat. Mycotoxin Res 31:137–143. https://doi.org/10.1007/s12550-015-0224-8

Phillips TW, Thoms EM, Demark J, Walse S (2012) Fumigation. In: Hagstrum DW, Phillips TW, Cuperus G (eds) Stored product protection, vol 14. Kansas State University, Manhattan, pp 157–178

Placinta CM, D'Mello CPF, MacDonald AMC (1999) A review of worldwide contamination of cereal grains and animal feed with *Fusarium mycotoxins*. Anim Feed Sci Technol 78:21–37

Preston GM (2004) Plant perceptions of plant growth-promoting Pseudomonas. Philos Trans R Soc Lond Ser B Biol Sci 359:907–918. https://doi.org/10.1098/rstb.2003.1384

Priyanka SR, Venkataramana M, Balakrishna K, Murali HS, Batra HV (2015) Development and evaluation of a multiplex PCR assay for simultaneous detection of major mycotoxigenic fungi from cereals. J Food Sci Technol 52:486–492. https://doi.org/10.1007/s13197-013-1001-3

Raghunathan AN, Muthu M, Majumder SK (1969) Control of internal fungi of sorghum by fumigation. J Stored Prod Res 5:389–392. https://doi.org/10.1016/0022-474X(69)90012-5

Sabillón L, Bianchini A (2016) From field to table: a review on the microbiological quality and safety of wheat-based products. Cereal Chem J 93:105–115. https://doi.org/10.1094/CCHEM-06-15-0126-RW

Sadhasivam S, Britzi M, Zakin V, Kostyukovsky M, Trostanetsky A, Quinn E, Sionov E (2017) Rapid detection and identification of mycotoxigenic fungi and mycotoxins in stored wheat grain. Toxins 9:302. https://doi.org/10.3390/toxins9100302

Schmidt R, Köberl M, Mostafa A, Ramadan EM, Monschein M, Jensen KB, Bauer R, Berg G (2014) Effects of bacterial inoculants on the indigenous microbiome and secondary metabolites of chamomile plants. Front Microbiol 5:64. https://doi.org/10.3389/fmicb.2014.00064

Sinha RN, Berck B, Wallace HAH (1967) Effect of phosphine on mites, insects, and microorganisms. J Econ Entomol 60:125–132. https://doi.org/10.1093/jee/60.1.125

Smith JE (1997) Aflatoxins. In: D'Mello JPF (ed) Plant and Fungal Toxicants, vol 8. CRC Press, Boca Raton, pp 269–285

Solanki MK, Abdelfattah A, Britzi M, Zakin V, Wisniewski M, Droby S, Sionov E (2019) Shifts in the composition of the microbiota of stored wheat grains in response to fumigation. Front Microbiol 10:1098. https://doi.org/10.3389/fmicb.2019.01098

Streit E, Naehrer K, Rodrigues I, Schatzmayr G (2013) Mycotoxin occurrence in feed and feed raw materials worldwide: long-term analysis with special focus on Europe and Asia. J Sci Food Agric 93:2892–2899. https://doi.org/10.1002/jsfa.6225

Suanthie Y, Cousin MA, Woloshuk CP (2009) Multiplex real-time PCR for detection and quantification of mycotoxigenic Aspergillus, Penicillium and Fusarium. J Stored Prod Res 45:139–145. https://doi.org/10.1016/J.JSPR.2008.12.001

Sylla J, Alsanius BW, Krüger E, Reineke A, Strohmeier S, Wohanka W (2013) Leaf microbiota of strawberries as affected by biological control agents. Phytopathology 103:1001–1011. https://doi.org/10.1094/PHYTO-01-13-0014-R

Tao A, Pang F, Huang S, Yu G, Li B, Wang T (2014) Characterisation of endophytic *Bacillus thuringiensis* strains isolated from wheat plants as biocontrol agents against wheat flag smut. Biocontrol Sci Tech 24:901–924. https://doi.org/10.1080/09583157.2014.904502

Wegulo SN (2012) Factors influencing deoxynivalenol accumulation in small grain cereals. Toxins 4:1157–1180. https://doi.org/10.3390/toxins4111157

Yuan QS, Yang P, Wu AB, Zuo DY, He WJ, Guo MW, Huang T, Li HP, Liao YC (2018) Variation in the microbiome, trichothecenes, and aflatoxins in stored wheat grains in Wuhan, China. Toxins 10:171. https://doi.org/10.3390/toxins10050171

Chapter 9
Progress on Chemical Management of Postharvest Diseases of Subtropical and Tropical Fruits

James E. Adaskaveg, Helga Förster, and Daniel Chen

Abstract Application of postharvest fungicides is a highly effective, targeted, and economical way to protect fruit from decay after harvest including during shipment and marketing with proper application methods, fungicide residues remain within maximum residue limits, similar to levels found after preharvest field applications. The United States have been on the forefront in developing these treatments. For sub-tropical crops such as citrus, pomegranates, pineapples, and avocados, new fungicides have been introduced to manage major diseases. They represent different FRAC (Fungicide Resistance Action Committee) groups ranging from conventional (propiconazole), reduced-risk (pyrimethanil; azoxystrobin; fludioxonil) to bio-pesticide (natamycin) compounds, and also include pre-mixtures (fludioxonil-propiconazole, -azoxystrobin, and -cyprodinil). Selection and development of these fungicides has been done gradually as products designated safer by the United States Environmental Protection Agency were identified that maintain high performance, use low rates, and have broad-spectrum activity similar to previous standards. Natamycin, a fermentation product of *Streptomyces* spp., is exempt from tolerance in North America, it potentially may be certified as an organic treatment, and resistance has not been reported in filamentous fungi in over 30 years of use. This may prove crucial in the management of *Penicillium* spp. and other postharvest decays of subtropical crops. Although moderately effective by itself, mixtures provide high efficacy and resistance management.

Keywords Postharvest fungicides · Chemical control · Bio-fungicides

J. E. Adaskaveg (✉) · H. Förster · D. Chen
Department of Microbiology and Plant Pathology, University of California, Riverside, CA, USA
e-mail: jim.adaskaveg@ucr.edu

D. Spadaro et al. (eds.), *Postharvest Pathology*, Plant Pathology in the 21st Century 11, https://doi.org/10.1007/978-3-030-56530-5_9

Introduction

The human population of the world continues to increase, and with changing demographics, 55% of the population is currently living in urban areas. Following this trend, by 2050, only 32% of humans will live in rural areas. Food demand will increase with population growth, including the demand for fresh fruit and vegetable commodities with high nutritional value as a growing number of societies will have improved standards of living. Prolonged storage and long-distance distribution of fresh produce has created new challenges in the management of fungal decays causing crop losses after harvest.

Most postharvest decays of perishable horticultural crops, including subtropical and tropical fruits and vegetables, are caused by fungi (Snowdon 1990, 1992). These fungi commonly disseminate themselves through the production of asexual spores (mitospores). Many fungi infect the host only through wounds such as *Penicillium* species on citrus and pineapple and *Ceratocystis* (*Thielaviopsis*) species on pineapple, banana, and other crops. Others, such as fungi in the Botryosphaeriaceace (e.g., *Lasiodiplodia theobromae*, *Phomopsis citri*), as well as *Colletotrichum* and *Phytophthora* species are well-known for their ability to infect fruit commodities directly through intact plant surfaces, causing both active and quiescent types of infection. Active infections will progress immediately and cause decay. In quiescent types of infection, the pathogen invades the outermost cell layers of the plant tissue but then remains inactive until the commodity ripens or the environment changes. Host-produced pectinases that soften fruit can activate the pathogen to cause decay or alternatively, the pathogen can modulate the pH around the infection site until the host tissue and micro-environment are conducive for enzymatic degradation (Prusky et al. 2001; Diéguez-Uribeondo et al. 2008).

The main subtropical and tropical fruit crops grown in the United States are avocado, citrus, mango, papaya, pineapple, and pomegranate. Examples of major postharvest decays occurring on these crops, their causal pathogens, as well as site and type of infection are shown in Table 9.1 and selected decays in Fig. 9.1. Different chemical strategies can be utilized to manage these types of infection.

Decay management depends on an integrated approach including cultural, preharvest, harvest, and postharvest practices that are essential components influencing the complex interaction between host, pathogen, and environment. For example, selection of crop variety, orchard practices, pest and disease management programs, as well as preharvest fungicides and their application timing can directly affect the potential of postharvest decay development. The type of preharvest fungicide, selected in respect to spectrum of activity, persistence, and systemic action, determines which type of infection present can be inhibited at harvest. Among postharvest practices, the use of fungicides is the most effective means to reduce decay (Eckert and Ogawa 1985; Adaskaveg et al. 2002), while residues remain within maximum limits, similar to levels found after preharvest field applications. Postharvest fungicide applications are targeted and economical, and may protect fruit from fungal attacks occurring before treatments, including quiescent

Table 9.1 Major postharvest decays in long-distance marketing of sub-tropical and tropical crops

Crop	Decay	Causal pathogen	Site of infection	Type of infection
Avocado	Anthracnose	*Colletotrichum gloeosporioides sensu lato*	Intact Surface	Quiescent or active
	Stem-end rots	*Lasiodiplodia theobromae*, *Phomopsis citri*, and other species	Stem end and injuries	Quiescent or active
Citrus	Green mold	*Penicillium digitatum*	Injury	Active
	Blue mold	*P. italicum*	Injury	Active
	Sour rot	*Geotrichum citri-aurantii*	Injury	Active
	Stem-end rots	*L. theobromae*, *P. citri*, and other species	Stem and blossom end	Quiescent or active
	Brown rot	*Phytophthora* spp.	Intact Surface	Quiescent or active
Pineapple	Black rot	*Ceratocystis paradoxa* (*Thielaviopsis paradoxa*)	Injury	Active
	Surface molds	*Penicillium*, *Cladosporium*, and other species	Injury	Active
Pomegranate	Gray mold	*Botrytis cinerea*	Injury and blossom end	Quiescent or active
	Alternaria rot	*Alternaria alternata*, *A. arborescens*	Flower?	Internal
	Aspergillus rot	*Aspergillus niger*	Flower?	Internal
Papaya	Anthracnose	*C. gloeosporioides s. lat.*	Intact Surface	Quiescent or active
	Black rot	*Stagonosporopsis caricae*	Injury	Active
	Stem-end rots	*L. theobromae*, *Diaporthe caricae-papayae*, *S. caricae*, *Phytophthora*, and other species	Stem and blossom end	Quiescent or active
	Phytophthora rot	*Phytophthora palmivora*, *P. tropicalis*	Intact Surface	Quiescent or active
	Rhizopus rot	*Rhizopus stolonifer*	Injury	Active
Mango	Anthracnose	*C. gloeosporioides s. lat.*	Intact Surface	Quiescent or active
	Alternaria rot	*A. alternata*	Injury	Active
	Stem-end rots	*L. theobromae*, *Phomopsis mangiferae*, *Botryosphaeria parva*, and others	Stem and blossom end	Quiescent or active

infections, and infections that are initiated during harvest and handling. They are also effective against infections occurring after treatment during postharvest handling, shipment, storage, and marketing. Hot water treatments are also effectively used on some tropical and subtropical crops such as mangos and papayas (Fallik 2010), and they are critical treatments when postharvest fungicides are not available or the final market restricts their use. However, these treatments do not have any residual efficacy against infections occurring after treatment. To overcome this issue

Fig. 9.1 Postharvest decays of sub-tropical and tropical fruits. (**a**) Green mold of orange; (**b**) Sour rot of lemon; (**c**) Anthracnose of avocado; (**d**) Penicillium surface mold and decay of pineapple; (**e**) Anthracnose of papaya; and (**f**) Gray mold of pomegranate

they can be used in combination with a fungicide for increased efficacy and residual activity. This strategy has been demonstrated using potassium phosphite for managing Phytophthora brown rot of citrus (Adaskaveg et al. 2015) and using selected fungicides for managing Penicillium decays of citrus (Smilanick et al. 1997, 1999). Biological control agents such as *Pseudomonas syringae* are registered as commercial products but do not provide the level of control needed for meeting successful marketing in distant locations.

Properties of Postharvest Fungicides for Managing Decays of Subtropical and Tropical Crops

Ideal characteristics of a postharvest fungicide include high activity against target pathogen(s), efficacy at low rates, broad spectrum of activity, and inhibition of new infections with persistent residues and existing infections, including those occurring during postharvest handling. Moreover, a postharvest fungicide should inhibit sporulation of the pathogen and production of secondary inoculum, an important aspect in citrus green mold management. This will prevent spoilage of fruit with spore masses and reduce the pathogen reservoir that may be selected for reduced sensitivity and ultimately resistance to the fungicide. Fungicides with locally systemic activity or translaminar movement into the host tissue are needed to inhibit quiescent infections that occur on many subtropical and tropical crops including avocado, mango, and papaya. Postharvest fungicides need to be environmentally non-toxic to non-target organisms and extremely safe to consumers and workers.

Moving Towards Safer Postharvest Fungicides

Postharvest treatments to preserve harvested fruits and vegetables have been used since antiquity. Elemental sulfur was one of the first chemicals applied. A multitude of chemicals were developed and introduced starting in the mid-1900s. Researchers in the United States have been on the forefront in developing these treatments. Older postharvest fungicides typically had a broad spectrum of activity and multiple sites of action, but many of them do not meet today's safety and high-efficacy standards. Still in wide use today on citrus fruits are the inorganic salts borax, sodium carbonate, and sodium bicarbonate (Palou et al. 2001). Use of the phenol sodium ortho-phenylphenate (SOPP, OPP), however, is currently minimal due to human health and environmental concerns.

More specific single-site mode of action fungicides that were registered in the 1970s and '80s, such as the methyl benzimidazole carbamate thiabendazole and the demethylation inhibitor imazalil, are also still in use on citrus and pome fruits or only citrus fruits, respectively in the United States. These fungicides, like many others that were registered prior to 1986, were required to go through re-registration in the United States. Some were renewed (e.g., thiabendazole, imazalil), whereas others were canceled (e.g., thiophanate-methyl, iprodione). These fungicides are now reviewed under risk eligibility decisions (REDs) by the United States Environmental Protection Agency (EPA) that apply new safety standards to these older fungicides. Additional labeling, re-formulation, personal protective equipment, or other requirements are often required to ensure the safety of workers and consumers.

In the development of new fungicides starting in the 1990s, emphasis was placed on minimizing human health risks and environmental toxicity (Adaskaveg and Förster 2010). Fungicides such as fludioxonil, pyrimethanil, and azoxystrobin that meet these standards are classified as 'reduced risk' by EPA and fit into integrated pest management approaches. They are registered in the United States for use on many fruits including some of the subtropical and tropical crops (Table 9.2). With mammalian toxicity concentrations (i.e., LD_{50} values) of more than 2000 mg/kg, these new treatments are the safest fungicides ever developed for postharvest use and are among chemicals most critically tested (Adaskaveg and Förster 2010). For relatively comparison, table salt, caffeine, and imazalil have oral LD_{50} values of 3000, 127, and 227 mg/kg of body weight, respectively. Fludioxonil was synthesized based on the natural product pyrrolnitrin, a secondary metabolite of *Burkholderia pyrrocinia*, whereas azoxystrobin was synthesized based on the natural products strobilurin-A and oudemansin A, secondary metabolites of *Strobilurus* and *Oudemansiella* species, respectively.

Reduced-risk concepts were extended in the United States with the introduction of the bio-pesticide category of disease control treatments that include natural products (e.g., microbial fermentation products, plant extracts) and minerals (e.g., sodium carbonate) as well as biological controls. Bio-pesticides are often exempt from tolerances because of their high safety and low mammalian toxicity, but they need to go through the same initial toxicology review process as conventional pesticides. Postharvest bio-fungicides recently registered are the natural products potassium phosphite and natamycin. The latter is a bio-fermentation product of *Streptomyces natalensis* and some other species of *Streptomyces* and has been used in the food industry as a preservative for decades. Another bio-pesticide being evaluated for postharvest use is polyoxin-D, a bio-fermentation product of *Streptomyces cacaoi* var. *asoensis*. This compound was recently approved as an organic pre- and postharvest treatment in the United States.

Postharvest Fungicide Registrations on Tropical and Subtropical Crops in the United States

Among tropical and subtropical fruits, an initial focus for development of postharvest fungicides was on citrus, a major crop that is known for its decay management challenges. Citrus is harvested almost year-round, fruit are sometimes stored for several months, and fungicide resistance in *Penicillium* species, the main decay pathogens, is common to previously registered fungicides (e.g., imazalil and thiabendazole). Newly registered fungicides represent different FRAC (Fungicide Resistance Action Committee) Codes and range from conventional (propiconazole - FRAC Code 3) to reduced-risk (pyrimethanil - FRAC Code 9; azoxystrobin - FRAC Code 11; fludioxonil - FRAC Code 12) and bio-pesticide (natamycin - FRAC Code 48; potassium phosphite – FRAC Code P07/33) compounds, and also include premixtures (imazalil - pyrimethanil, fludioxonil-azoxystrobin) (Table 9.2). These products were selected and developed gradually as safer compounds were identified that still maintain high performance. They are used at low rates and have

Table 9.2 Postharvest fungicides for citrus in the United States, their regulatory and FRAC classification, efficacy rating, and resistance risk and type

Fungicides and pre-mixtures	Regulatory classification[b]	FRAC Code	Efficacy ratings in decay management[a]				Resistance risk and type (when known)
			Green Mold decay	Green mold anti-sporulation	Sour rot	Brown rot	
Thiabendazole (TBZ)	C	1	+++/+	+++	–	–	*Penicillium* spp.: high – qualitative
Imazalil	C	3	++++/++	++++	–	–	*Penicillium* spp.: high – quantitative
Inorganic salts (borax, sodium carbonate, and sodium bicarbonate)	BP	–	++	–	++	–	No resistance
Azoxystrobin	RR	11	+++/(+)	+++	–	–	*Penicillium* spp.: moderate, no field resistance, likely qualitative
Fludioxonil	RR	12	++++/ (+++)	+++	–	–	*Penicillium* spp.: moderate, no field resistance, natural resistance unfit
Natamycin	BP	49	+++	–	+++	–	*Penicillium* and *Geotrichum* spp.: low, no resistance ever found in filamentous fungi
Potassium phosphite	BP	P07/33	+	–	–	+++/+	*Phytophthora* spp.: moderate, resistance detected

(continued)

Table 9.2 (continued)

Fungicides and pre-mixtures	Regulatory classification[b]	FRAC Code	Efficacy ratings in decay management[a]				Resistance risk and type (when known)
			Green Mold decay	Green mold anti-sporulation	Sour rot	Brown rot	
Propiconazole	C	3	++++/++	+++	++++	–	*Penicillium* and *Geotrichum* spp.: high – quantitative
Pyrimethanil	RR	9	++++/+	+/–	–	–	*Penicillium* spp.: high – qualitative
Imazalil + pyrimethanil	RR + C	3 + 9	++++/++	++++	–	–	*Penicillium* spp.: high – quantitative and qualitative, respectively
Fludioxonil + azoxystrobin	RR + RR	11 + 12	++++	++++	–	–	*Penicillium* spp.: moderate, no field resistance

[a]Rating scale for efficacy: ++++ = excellent and consistent; +++ = very good and reliable; ++ = moderate and variable; + = mimimal and sometimes ineffective; – = not effective. Ratings separated by a slash refer to decay caused by isolates sensitive/resistant to the fungicide. Ratings in parentheses refer to the efficacy of the fungicide in managing specific decays caused by laboratory-selected, resistant isolates

[b]Regulatory classifications: *C* conventional, *RR* reduced risk, *BP* bio-pesticide

broad-spectrum activity, similar to previous standards. Although the triazole propiconazole belongs to the same FRAC Code 3 as the imidazole imazalil (i.e., the demethylation inhibitors – DMIs) and there is cross-resistance between the two, propiconazole has much more favorable human safety characteristics. It was selected because it is highly active against sour rot and is the first highly effective residual fungicide registered against this decay (Fig. 9.2). Natamycin is the second most effective fungicide against sour rot and is also effective against green mold (Fig. 9.2).

Imazalil, propiconazole, as well as the methyl-benzimidazole carbamate (MBC) thiabendazole are highly inhibitory to wild-type sensitive (S) isolates of *Penicillium digitatum*. However, resistant isolates (R) to MBC and DMI fungicides are common in citrus packinghouses. Fludioxonil, azoxystrobin, and pyrimethanil are very effective against decays caused by $DMI^{S,R}$ and $MBC^{S,R}$ isolates of *Penicillium* spp., whereas natamycin is moderately effective against Penicillium decays and sour rot.

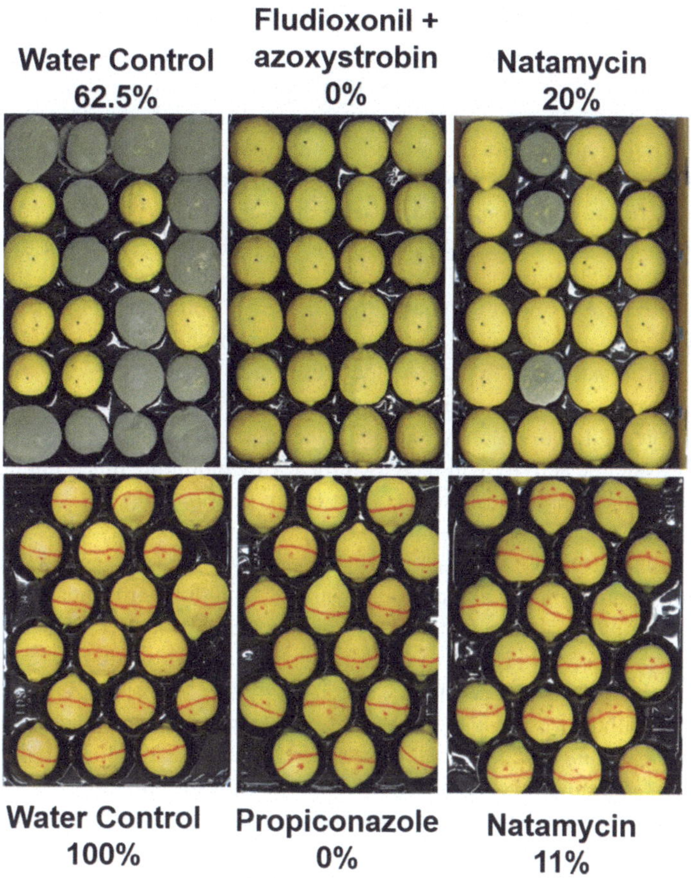

Fig. 9.2 Efficacy of postharvest fungicide treatments for managing green mold (top row) and sour rot (bottom row) of Eureka lemons in a commercial packing line study. Fruit were marked with a black dot or red line, wound-inoculated with spores of *Penicillium digitatum* or *Geotrichum citri-aurantii*, respectively, placed on the packingline among commercial fruit, and recovered after treatment. Incidence (%) of each decay is indicated under each treatment

These compounds vary in their risk to develop resistance in pathogen populations ranging from low (i.e., natamycin), moderate (i.e., azoxystrobin, fludioxonil), to high (i.e., propiconazole, pyrimethanil) (Table 9.2). The risk assessments are based on *in vitro* assays (Kanetis et al. 2010; McKay et al. 2012) and observations in packinghouses after prolonged usage. A unique aspect of natamycin is that resistance has never been reported in filamentous fungi despite extensive use in the food industry for more than 30 years. This may prove crucial in the management of Penicillium and other postharvest decays of subtropical crops when natamycin is used as a mixture partner in an anti-resistance strategy (Adaskaveg et al. 2019). Currently, pre-mixture treatments of fungicides belonging to different FRAC Codes are a recommended practice. Pre-mixtures also broaden the spectrum of activity of a treatment. In addition to its efficacy against decay pathogens and anti-resistance properties, natamycin has the potential to be certified as an organic treatment. The bio-fungicide potassium phosphite is the first effective compound against citrus brown rot caused by *Phytophthora* species.

The current selection of postharvest fungicide treatments for tropical and subtropical fruit crops other than citrus is much more limited, however, new registrations are pending and others are in development (Table 9.3). The development of new treatments for fruit crops such as avocado, mango, and papaya where field infections of decay pathogens often become quiescent should be a priority. This is because fruit that are healthy-appearing at harvest and during packing may arrive decayed at the market or have a short shelf life. To inhibit quiescent infections, postharvest treatments need to be able to penetrate into the outermost fruit layers. Efficacy of the treatment may be increased in combination with a heated water treatment if the crop is tolerant to these temperatures.

On avocado, the currently registered fludioxonil is very effective against all major decay pathogens (Table 9.3), however, the pending registration of azoxystrobin that has locally systemic activity will increase efficacy by inhibiting quiescent infections of *Colletotrichum* spp. and the stem end rot fungi. Fludioxonil is also permitted for postharvest use on mango and papaya, and its spectrum of activity covers the major postharvest pathogens but lacks activity against quiescent infections. An older DMI compound, triadimefon, as well as fludioxonil and a fludioxonil-propiconazole premixture are available on pineapple. Fludioxonil is weakly effective against black rot that can be managed using the two DMI compounds. Fludioxonil, however, is very effective against surface mold and decay caused by DMI-resistant *Penicillium* spp. populations. Although surface mold is more of a visual problem and does not affect the internal fruit tissues, it leads to consumer rejection at the final market. Only fludioxonil is currently registered on pomegranate, but a pre-mixture with cyprodinil is pending registration. This pre-mixture is being introduced to prevent resistance development in *Botrytis cinerea*, the major decay pathogen of pomegranate. Although fludioxonil has high intrinsic activity against *Alternaria* species, it is not effective against Alternaria rot of pomegranate. This disease develops internally in the fruit, mostly before harvest, from infections in the field going as far back as flowering and cannot be managed by any postharvest treatment.

Conclusion

New effective postharvest fungicide treatments have become available that target major decay pathogens of subtropical and tropical fruits, and additional ones are in development. They have great potential to improve local and long-distance distribution and make decay-free, healthy produce available to the growing world population. Fungicide registrations in the United States over the past 25 years progressed from compounds classified as conventional to reduced-risk, and most recently, to bio-pesticides, reflecting increased standards in human and environmental safety. Future registrations will include additional pre-mixtures to increase the spectrum of activity and reduce the risk of resistance development. A challenge still remains to educate consumers that these postharvest treatments undergo the same or even more stringent safety evaluations as

Table 9.3 Postharvest fungicides for selected tropical and subtropical fruits in the United States, their regulatory classification and efficacy rating

			Efficacy ratings in decay management[a]							
Crop	Fungicide[b]	Regulatory classification[c]	Black rot	Anthracnose	Gray mold	Stem-end rots	Rhizopus rot	Alternaria rot	Cladosporium rot/ surface mold	Penicillium decays/ surface mold
Avocado	Azoxystrobin	RR	NA[d]	+++	NA	+++	+++	NA	NA	NA
	Fludioxonil	RR	NA	++++	NA	+++	++++	NA	NA	NA
	Propiconazole (pending)	C	NA	+++	NA	+++	+++	NA	NA	NA
Mango	Fludioxonil	RR	NA	+++	++++	+++	+++	++++	++++	++++
Papaya	Fludioxonil	RR	NA	+++	++++	+++	+++	++++	++++	++++
Pineapple	Fludioxonil	RR	+	NA	NA	NA	NA	NA	++++	++++
	Natamycin	BP	+	NA	NA	NA	NA	NA	++++	++++
	Triadimefon	C	+++	NA	NA	NA	NA	NA	++++	++++
	Fludioxonil + propiconazole	RR + C	+++	NA	NA	NA	NA	NA	++++	++++
Pomegranate	Fludioxonil	RR	NA	NA	++++	NA	+++	–	++++	++++
	Fludioxonil + cyprodinil (pending)	RR	NA	NA	++++	NA	+++	–	++++	++++

[a]Rating scale for efficacy: ++++ = excellent and consistent; +++ = very good and reliable; ++ = moderate and variable; + = mimimal and sometimes ineffective; – = not effective

[b]'Pending' indicates that the fungicide is in the registration process

[c]Regulatory classifications: *C* conventional, *RR* reduced risk, *BP* bio-pesticide

[d]*NA* the decay is not of importance on the crop

any other pesticide applied to the growing crop in the field and that the same maximum residue limits (MRLs) apply. Moreover, with oral LD_{50} values of more than 2000 mg/kg body weight, reduced-risk fungicides have similar toxicity profiles as bio-fungicides and possibly many natural products that are proposed as postharvest treatments, although they are not chemically well-characterized. An important additional step for the new registrations is to continue to establish and harmonize national and international MRLs that will facilitate international trade of these crops.

References

Adaskaveg JE, Förster H (2010) New developments in postharvest fungicide registrations for edible horticultural crops and use strategies in the United States. In: Prusky D, Gullino ML (eds) Post-harvest pathology. Plant pathology in the 21th century: contributions to the 9th international congress. Springer, Dordrecht/Heidelberg/London/New York, pp 107–117

Adaskaveg JE, Förster H, Sommer NF (2002) Principles of postharvest pathology and management of decays of edible horticultural crops. In: Kader A (ed) Postharvest technology of horticultural crops, DANR Publication 3311, 4th edn. University of California, Oakland, pp 163–195

Adaskaveg JE, Hao W, Förster H (2015) Postharvest strategies for managing Phytophthora brown rot of citrus using potassium phosphite in combination with heat treatments. Plant Dis 99:1477–1482

Adaskaveg JE, Förster H, Chen D (2019) Positioning natamycin as a post-harvest fungicide for citrus. Citrograph 10(4):62–65

Diéguez-Uribeondo J, Förster H, Adaskaveg JE (2008) Visualization of localized pathogen-induced pH modulation in almond tissues infected by *Colletotrichum acutatum* using confocal scanning laser microscopy. Phytopathology 98:1171–1178

Eckert JW, Ogawa JM (1985) The chemical control of postharvest diseases: subtropical and tropical fruits. Annu Rev Phytopathol 23:421–454

Fallik E (2010) Hot water treatments of fruits and vegetables for postharvest storage. Hortic Rev 38:191–212

Kanetis L, Förster H, Adaskaveg JE (2010) Determination of natural resistance frequencies in *Penicillium digitatum* using a new air-sampling method and characterization of fludioxonil- and pyrimethanil-resistant isolates. Phytopathology 100:738–746

McKay AH, Förster H, Adaskaveg JE (2012) Toxicity of selected fungicides to the postharvest pathogens *Galactomyces citri-aurantii*, *G. geotrichum*, and *Penicillium digitatum* and resistance potential to propiconazole. Plant Dis 96:87–96

Palou L, Smilanick JL, Usall J, Vinas I (2001) Control of postharvest blue and green molds of oranges by hot water, sodium carbonate, and sodium bicarbonate. Plant Dis 85:371–376

Prusky D, McEvoy JL, Leverentz B, Conway WS (2001) Local modulation of host pH by *Colletotrichum* species as a mechanism to increase virulence. Mol Plant-Microbe Interact 14:1105–1113

Smilanick JL, Mackey BE, Reese R, Usall J, Margosan DA (1997) Influence of concentration of soda ash, temperature, and immersion period on the control of postharvest green mold of oranges. Plant Dis 81:379–382

Smilanick JL, Margosan DA, Mlikota F, Usall J, Michael IF (1999) Control of citrus green mold by carbonate and bicarbonate salts and the influence of commercial postharvest practices on their efficacy. Plant Dis 83:139–145

Snowdon AL (1990) A color atlas of postharvest diseases and disorders of fruits and vegetables. Vol. 1: General introduction and fruits. CRC Press, Boca Raton. 302 pp

Snowdon AL (1992) A color atlas of postharvest diseases and disorders of fruits and vegetables, Vegetables, vol 2. CRC Press, Boca Raton, p 416

Chapter 10
Integration of Postharvest Fungicides and Fruit Sanitation Treatments to Optimize Decay Control and Address Food Safety Concerns

James E. Adaskaveg, Helga Förster, Daniel Chen, and Kevin A. Nguyen

Abstract Postharvest fungicide applications with aqueous recycling drenches are one on the most effective strategies to prevent fruit decay. Lower rates are used, and fungicide residues in fruit are generally lower than when using other application strategies. With the establishment of the Food Safety Modernization Act (FSMA) in the United States, sanitation of recycling fungicide drenches is mandatory due to the potential contamination with food-borne human bacterial pathogens. Thiabendazole, fludioxonil, azoxystrobin, and difenoconazole are compatible with the sanitizer sodium hypochlorite; whereas imazalil, pyrimethanil, and natamycin are not. All of these fungicides except natamycin are compatible with peroxyacetic acid. The use of recycling fungicide drench treatments is jeopardized due to this incompatibility of sanitizers with fungicides and additionally, because of volatile odors of sanitizers that may affect packinghouse workers. Acidification (pH 3.5) with natural organic acids (e.g., citric acid) and food-grade surfactants is compatible with postharvest fungicides evaluated, non-odoriferous, non-phytotoxic, and highly toxic to human bacterial pathogens. Sanitation of recycling fungicide drenches by acidification meets FSMA requirements and allows highly effective decay control but low pH and concentration need to be maintained continuously to be effective.

Keywords Postharvest fungicides · Sanitizers · Compatibility

J. E. Adaskaveg (✉) · H. Förster · D. Chen · K. A. Nguyen
Department of Microbiology and Plant Pathology, University of California, Riverside, CA, USA
e-mail: jim.adaskaveg@ucr.edu

D. Spadaro et al. (eds.), *Postharvest Pathology*, Plant Pathology in the 21st Century 11, https://doi.org/10.1007/978-3-030-56530-5_10

Introduction

Under the United States Food Safety Modernization Act (FSMA) that was signed into law in 2011, fruit packinghouses must sanitize recirculating fungicide solutions to ensure that human bacterial pathogens do not accumulate and cause food-borne illnesses to consumers. The overall goal of FSMA is to protect public health and reduce food-borne illnesses by strengthening the food safety system. To prevent or minimize potential hazards, the rules implement a preventive framework where safety standards are established by the United States Food and Drug Administration (FDA) to produce safety and preventative measures for human food. The food industries in the United States are responsible that standards are met through accreditation by third-party auditors for domestic food production and by a separate verification program for foreign suppliers.

Application of Postharvest Fungicides for Managing Fruit Decays

Postharvest fungicide applications to fruit crops can be divided into low- and high-volume systems (Fig. 10.1) based on the volume of fungicide solution used per tonnage of fruit, while the same amount of active ingredient is applied per total fruit weight (Adaskaveg et al. 2002). Low-volume systems such as controlled droplet applicators (CDAs), air-nozzle, or wig-wag systems use 30 to 130 liters per 100,000 kg of fruit, whereas high-volume systems such as T-Jet sprays, flooders, drenchers, or dip tanks use a minimum of 250 to 450 liters per 100,000 kg fruit. Low-volume systems have an advantage with no disposal issues because all of the prepared solutions are applied to the fruit.

These systems often use high concentrations of a postharvest fungicide in a fruit coating (i.e., wax) and rely on moving brushes to distribute it over the fruit surface with little to no run-off. High-volume application systems outperform low-volume systems in decay control efficacy because they provide better fungicide coverage (Fig. 10.2; Förster et al. 2007; Kanetis et al. 2008). Lower concentrations of fungicides in aqueous suspensions or solutions are applied that evenly coat the entire fruit surface and are deposited into micro-injuries. The result is high efficacy, lower use rates, and reduced fungicide residues on fruit. Sporulation control, however, is optimized when fungicides are used in fruit coatings that also improve fruit appearance and reduce water loss during storage and marketing. Thus, staged sequential postharvest treatments with aqueous fungicide followed by fruit coating applications result in the highest level of decay management. Low- and high-volume systems can be used economically but eventually, there are left-over fungicide solutions in any high-volume system that will need to be periodically disposed. For the last 15 years in citrus and stone fruit (e.g., cherry, peach, plum) packing facilities in California, postharvest fungicide recycling in-line drench or flooder application

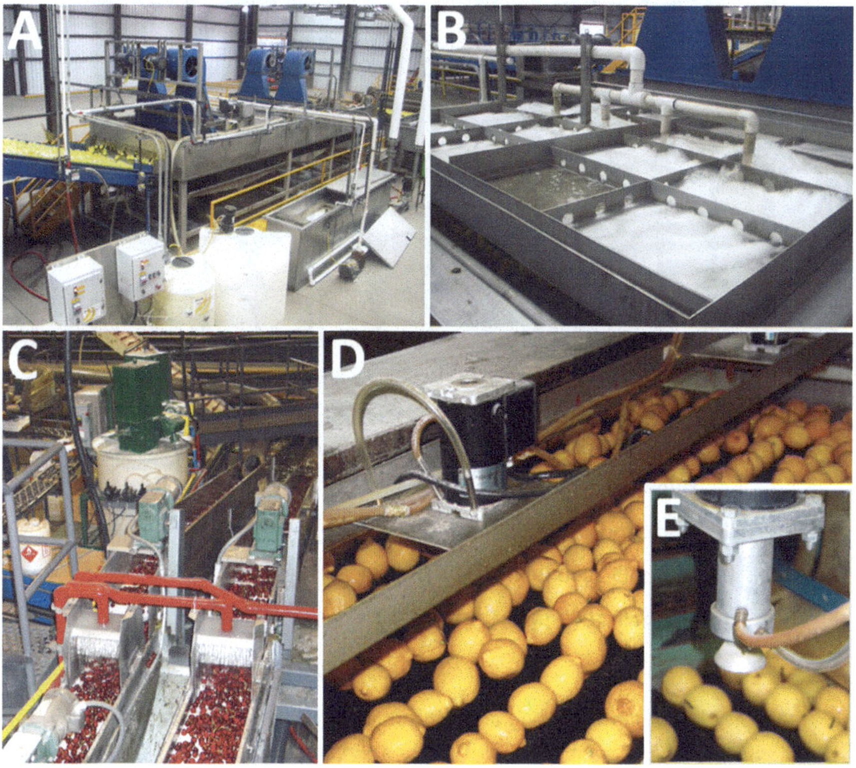

Fig. 10.1 High- (**a**, **b**, **c**) and low- (**d**, **e**) volume systems used for postharvest fungicide applications to fruit crops. Lemons are treated in (**a**, **b**, **d**, and **e**); whereas cherry fruit are treated in (**c**)

systems have become widely used for achieving high decay control. For some fungicides, including most of the newer reduced-risk compounds (Adaskaveg and Förster 2010), waste solutions from recycling application systems can be disposed of legally through the sewer system or to holding ponds, or they can be used for dust control on orchard roads. Other more toxic fungicides (e.g., sodium orthophenylphenate, SOPP, or orthophenylphenol, OPP) need to be disposed of by specialized companies that de-toxify industrial chemicals.

Sanitizing Postharvest Fungicides in Recycling High-Volume Application Systems Any aqueous solution that is in contact with environmental samples (e.g., fruit on a packing line) and that is kept for a prolonged period at ambient temperatures is subject to microbial contamination. Contaminants of high-volume, recycling, aqueous fungicide solutions may include innocuous environmental bacteria and fungi as well as plant pathogens, but of most concern are bacteria causing human diseases. Bacterial pathogens such as *Cyclospora cayetanensis, Escherichia coli* O157:H7, *Listeria monocytogenes*, *Salmonella* spp., and *Shigella* spp., as well as viruses such as Hepatitis A and Norovirus can contaminate raw (i.e., fresh)

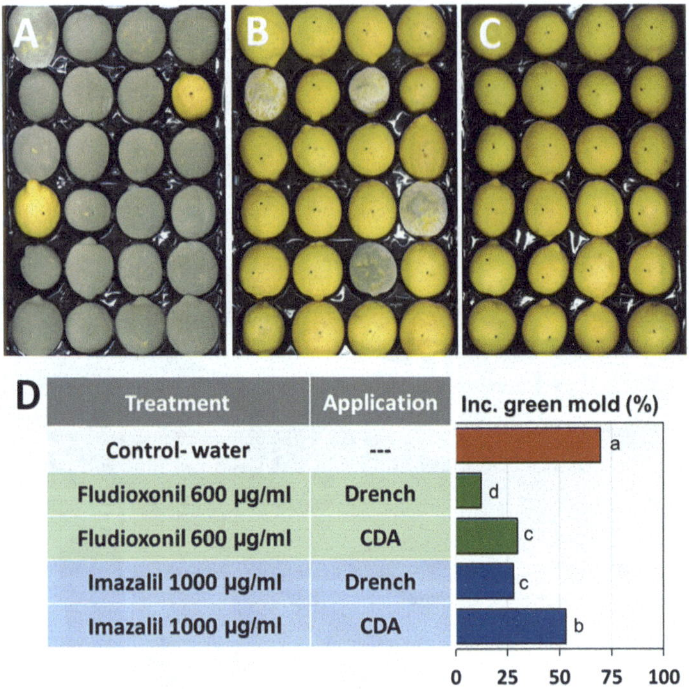

Fig. 10.2 Efficacy of fludioxonil in low and high-volume fungicide application systems for managing green mold (*Penicillium digitatum*) of lemons. In the two application systems, the same amount of fludioxonil per weight of fruit was used (i.e., 66 g/10,000 kg). (**a**) Untreated control fruit; (**b**) Fruit treated by low-volume, controlled droplet application (CDA) in a packing fruit coating; (**c**) Fruit treated by high-volume fungicide drench application followed by a CDA application with fruit coating without fungicide; and (**d**) incidence of green mold in a study comparing low- and high-volume applications of fludioxonil and imazalil

agricultural products (Akbas and Cag 2016). Although fruit are always washed with a sanitizing agent prior to exposure to re-cycling fungicide solutions, microbial contamination can still occur because small areas of fruit surfaces may escape the cleaning process, and oxidizing sanitizers can be neutralized in the presence of organic compounds (Teng et al. 2018). Additionally, there may be sources of re-contamination if solutions are kept for extended periods. To meet safety standards, fungicide solutions have to be effectively sanitized using oxidizers such as sodium or calcium hypochlorite or peroxyacetic acid (PAA). Unfortunately, some fungicides are incompatible with these sanitizers. The active ingredient breaks down, and subsequently, performance of the fungicide is lost. Imazalil, natamycin, and pyrimethanil are incompatible with sodium hypochlorite, whereas natamycin is also incompatible with PAA. Additionally, these sanitizers at concentrations that are effective in killing human bacterial pathogens may have odors that are offensive to workers in the immediate area.

Alternative Sanitizers

Sanitizers for recirculating fungicide systems need to be: (1) compatible with fungicides; (2) not phytotoxic to the commodity; (3) odor-free and non-irritant to workers; (4) fast-acting and kill human pathogens within a short time period of ideally less than 30 s (although a 4–5 min dwell time may be acceptable); and (5) should cause a > 5-log-reduction in colony forming units (cfu)/ml (i.e., a 99.999% reduction). A food-grade organic acid, citric acid (CA), was evaluated as an alternative and odorless sanitizer. CA is designated 'Generally Recognized as Safe' or GRAS in the United States. Commercial formulations of organic acids including CA and lactic acid registered for food use have shown high toxicity against most food-borne human bacterial pathogens including *E. coli* O157:H7 and pathogenic strains of *Listeria, Salmonella, Pseudomonas, Shigella, Vibrio*, and *Yersinia* species (Akbas and Ölmez 2007; Akbas and Cag 2016; Yoon and Lee 2018).

Laboratory Evaluation of CA as a Sanitizer of Fungicide Solutions

CA was tested *in vitro* against a non-pathogenic surrogate strain of *E. coli* and against *Pseudomonas syringae*, a surrogate for *P. aeruginosa*. In direct exposure assays, bacteria-contaminated solutions of azoxystrobin or fludioxonil, two currently widely used citrus postharvest fungicides with different modes of action (Adaskaveg and Förster 2010), as well as of natamycin, a newly registered postharvest biofungicide for citrus (Adaskaveg et al. 2019), were exposed to 1100 μg/ml CA with or without 60 μg/ml of the food-grade surfactant sodium dodecylbenzenesulfonate (SDBS) for 60 s at 48 °C and then plated onto agar media. Viability of bacteria was assessed based on colony counts. Log_{10}-reductions in *E. coli* colony counts were compared to a water control and are shown for the three fungicides in Table 10.1.

Using CA by itself without SDBS, a 5-$\log_{10}$ reduction was only obtained using fludioxonil. The addition of SDBS significantly improved the toxicity of CA against *E. coli* for azoxystrobin and natamycin mixtures but not for fludioxonil. Levels of sanitation achieved comply with FSMA regulations for sufficient sanitation efficacy for the three fungicide solutions. Exposure of *E. coli* and *P. syringae* to fungicide solutions amended with 1100 μg/ml CA and 60 μg/ml SDBS inactivated all or nearly all cells within 120 s at ambient-temperature and within 60 s when solutions were heated to 48 °C (data not shown). In all tests conducted, *P. syringae* was found to be more sensitive to acidification treatments and therefore, subsequent studies were conducted using *E. coli*. Furthermore, the bactericidal activity was maintained when solutions were re-contaminated with either bacterial species after 24 h. These results demonstrate that a heated or ambient-temperature fungicide solution acidified with citric acid-SDBS is highly effective in killing bacteria within short time

Table 10.1 *In vitro* toxicity of citric acid with and without a surfactant against a non-pathogenic strain of *E. coli* in the presence of selected postharvest fungicides at 48°C[a]

				Reduction in *E. coli* cfu ($\log_{10}$)[b]					
				Azoxystrobin		Fludioxonil		Natamycin	
Sanitizer	pH range	Exposure time (s)	Surfactant	300 µg/ml	LSD	300 µg/ml	LSD	1000 µg/ml	LSD
Citric acid 1100 µg/ml	2.74–2.83	60	–	2.26	A	5.29	A	1.14	A
			+	6.40	B	5.35	A	6.46	B

[a]Aqueous suspensions of azoxystrobin, fludioxonil, or natamycin were prepared with 1100 µg/ml citric acid with or without 60 µg/ml of the food-grade surfactant sodium dodecylbenzenesulfonate (SDBS). Suspensions were heated to 48 °C, *E. coli* (2×10^6 cfu/ml final concentration) was added, aliquots were diluted after 60 s exposure time and were then plated onto agar media. Viability of bacteria was assessed based on colony counts after 2 days

[b]Sanitation is regarded as a reduction in microbial contamination by 5-$\log_{10}$ or more

periods and that this method is suitable for sanitizing fungicide solutions in recycling reservoirs.

Fruit Treatments with Citric Acid-Fungicide Solutions for Managing Green Mold

Studies were conducted using lemon fruit inoculated with *P. digitatum* and treated with acidified (pH 3.5 to 3.75; SDBS added) or non-acidified solutions of azoxystrobin, fludioxonil, or natamycin prepared 24 h before use. For treatments, fungicide solutions were contaminated with *E. coli* (2 x 10^6 cfu final concentration), heated (48 °C) in a recirculation flooder tank of an experimental packingline, and applied to fruit by drenches. There was no significant difference in decay control efficacy for each fungicide using both types of solutions and therefore, acidification did not interfere with fungicide activity (Table 10.2). The incidence of green mold was reduced by 100% using azoxystrobin or fludioxonil or by 74% using natamycin as compared with the untreated control where 56.3% of the fruit developed green mold. These levels of decay control are typical for the three fungicides. The biofungicide natamycin is generally less effective against green mold than the other fungicides, but its strengths are activity against sour rot and a very low potential for resistance development, characteristics that favor its use as a mixture partner for other fungicides (Adaskaveg et al. 2019).

In the experiment described above, fungicide solutions after treatment of fruit were plated onto agar media to determine the viability of the *E. coli* inoculum. Bacterial populations were reduced in size by 2.37- to 2.61-$\log_{10}$ as compared with initial counts when fungicide solutions were not amended with CA and SDBS (Table 10.2). This reduction can be explained by exposure to heat (48 °C), because bacteria were disrupted by passing through the centrifugal pump that was used to

Table 10.2 Efficacy of selected postharvest fungicides by themselves or mixed with citric acid in managing green mold of inoculated lemon fruit on an experimental packingline line and toxicity of fungicide and fungicide-sanitizer suspensions against a non-pathogenic strain of *E. coli*

	% Reduction in decay incidence (compared to control)[a]						Reduction in *E. coli* cfu (log_{10})[b]					
	Azoxystrobin		Fludioxonil		Natamycin		Azoxystrobin		Fludioxonil		Natamycin	
Sanitizer	300 μg/ml	LSD	300 μg/ml	LSD	1000 μg/ml	LSD	300 μg/ml	LSD	300 μg/ml	LSD	1000 μg/ml	LSD
None (fungicide only)	96.3	A	92.5	A	85.3	A	2.39	A	2.61	A	2.37	A
Citric acid 1000 μg/ml + SDBS 55 μg/ml	100	A	100	A	74.1	A	6.30	B	6.30	B	2.39	A

[a]Aqueous suspensions of azoxystrobin, fludioxonil, or natamycin were exposed to citric acid (pH < 3.75) with 55 μg/ml sodium dodecylbenzenesulfonate (SDBS) for 24 h. Suspensions were heated to 48 °C in a recirculating flooder tank and used to treat fruit inoculated with *P. digitatum*. E. coli (2×10^6 cfu/ml final concentration) was added, aliquots were diluted after 120 s exposure time and were then plated onto agar media. Viability of bacteria was assessed based on colony counts after 2 days

[b]Sanitation is regarded as a reduction in microbial contamination by 5-log_{10} or more

recirculate the fungicide solution, or possibly by a combination of both. When fungicide solutions were amended with the sanitizer and the detergent, a reduction in bacterial counts by 6.3-$\log_{10}$ was obtained using azoxystrobin or fludioxonil, however, not when using natamycin. The incubation time for the bacteria in the packing-line study was approximately twice as long (i.e., 120 s) as compared with the laboratory study (Table 10.1), but in the presence of fruit, longer exposure times may be needed to reduce bacterial populations when using natamycin. In studies not presented here, *E. coli* cells were inactivated within 1 h in natamycin-CA-SDBS solutions that were used for treating fruit. This time period is not unrealistic under commercial conditions, where heated solutions can be continuously exposed to acidic conditions of the sanitizer during the day and overnight. This suggests that citric acid can be an alternative sanitizer for the fungicides evaluated. No phytotoxicity was observed on fruit with any of the treatments.

Conclusion

Under the FSMA, fruit packinghouses must sanitize recirculating fungicide solutions to ensure that human bacterial pathogens do not accumulate and cause food-borne illnesses to consumers. Typically, packinghouses use oxidizers such as sodium hypochlorite or peroxyacetic acid for sanitation. These compounds may be incompatible with certain fungicides and produce odors that may irritate workers. To identify alternative, highly effective, and odorless sanitizers, we tested CA that is designated as exempt from tolerance in the United States. This organic acid and its commercial formulations are registered for food use and have high toxicity against most food-borne human bacterial pathogens. Using surrogates for bacterial pathogens, we demonstrate that CA can be effectively used to sanitize recirculating fungicide solutions when the pH is <3.75 without loss of decay control activity and without phytotoxicity to fruit. Thus, acidification can be a highly effective alternative sanitation method to oxidation for recirculating fungicide solutions. Other organic acids such as lactic acid may function similarly to CA, can be easily tested in selected fungicide solutions, and are commercially available.

References

Adaskaveg JE, Förster H (2010) New developments in postharvest fungicide registrations for edible horticultural crops and use strategies in the United States. In: Prusky D, Gullino ML (eds) Post-harvest pathology. Plant pathology in the 21th century: contributions to the 9th international congress. Springer, Dordrecht/Heidelberg/London/New York, pp 107–117

Adaskaveg JE, Förster H, Chen D (2019) Positioning natamycin as a post-harvest fungicide for citrus. Citrograph 10:62–65

Adaskaveg JE, Förster H, Sommer NF (2002) Principles of postharvest pathology and management of decays of edible horticultural crops. In: Kader A (ed) Postharvest technology of horticultural crops, DANR Publication 3311, 4th edn. University of California, Oakland, pp 163–195

Akbas MY, Cag S (2016) Use of organic acids for prevention and removal of *Bacillus subtilis* biofilms on food contact surfaces. Food Sci Technol Int 22:587–597

Akbas MY, Ölmez H (2007) Inactivation of *Escherichia coli* and *Listeria monocytogenes* on iceberg lettuce by dip wash treatments with organic acids. Lett Appl Microbiol 44:619–624

Förster H, Driever GF, Thompson DC, Adaskaveg JE (2007) Postharvest decay management for stone fruit crops in California using the "reduced-risk" fungicides fludioxonil and fenhexamid. Plant Dis 91:209–215

Kanetis L, Förster H, Adaskaveg JE (2008) Optimizing efficacy of new postharvest fungicides and evaluation of sanitizing agents for managing citrus green mold. Plant Dis 92:261–269

Teng Z, van Haute S, Zhou B, Hapeman CJ, Millner PD, Wang O, Luo Y (2018) Impacts and interactions of organic compounds with chlorine sanitizer in recirculated and reused produce processing water. Plos ONE. https://doi.org/10.1371/journal.pone.0208945

Yoon JH, Lee SY (2018) Review: comparison of the effectiveness of decontaminating strategies for fresh fruits and vegetables and related limitations. Crit Rev Food Sci Nutr 58:3189–3208

Chapter 11
GRAS Salts as Alternative Low-Toxicity Chemicals for Postharvest Preservation of Fresh Horticultural Products

Lluís Palou and María B. Pérez-Gago

Abstract Among means alternative to chemical fungicides to control postharvest decay of fresh horticultural products, low-toxicity chemicals, classified as food additives or Generally Recognized As Safe (GRAS) compounds, are of interest because of their low toxicological effects on mammals and minimal impact on the environment. These chemicals include some essential oils, plant extracts, and other natural compounds, but also synthetic inorganic or organic salts such as carbonates, sorbates, benzoates, acetates, paraben salts, silicates, etc. Major advantages of these salts are their inherent antimicrobial activity and high solubility in water. In general, research with GRAS salts initiates with the *in vitro* evaluation of the antimicrobial activity of aqueous solutions against target postharvest pathogens. Selected salt solutions and concentrations are then assayed in *in vivo* trials, first at the laboratory scale with fresh produce artificially inoculated with the pathogen and afterwards at semi-commercial or commercial scale with naturally infected produce. GRAS salts can also be used as antimicrobial ingredients of synthetic composite edible coatings, which can have an impact on both physiological and pathological factors limiting the postharvest life of fresh horticultural products. Coating formulation and selection is based on *in vivo* decay control ability and overall produce quality maintenance during cold storage.

Keywords Fresh produce · Postharvest decay · Fungicide-free control · Food preservatives · Antimicrobial edible coatings

L. Palou (✉) · M. B. Pérez-Gago
Centre de Tecnologia Postcollita (CTP), Institut Valencià d'Investigacions Agràries (IVIA), València, Spain
e-mail: palou_llu@gva.es

D. Spadaro et al. (eds.), *Postharvest Pathology*, Plant Pathology in the 21st Century 11, https://doi.org/10.1007/978-3-030-56530-5_11

Introduction

Considerable losses of fresh horticultural products worldwide are due to physiological and pathological issues affecting postharvest life. Product storability and shelf life can be dramatically reduced by microbial pathogens causing postharvest decay. Typically, fungi are the most prevalent causal agents, particularly in the case of fresh fruits. In general, fungal infections that occur during bloom or fruit development in the field but develop after harvest are known as latent or quiescent infections, and infections that occur on mature fruit near, during, or after harvest through peel wounds or injuries are known as wound infections. Both types of infections can cause substantial losses of the most important horticultural commodities. For instance, worldwide economically important postharvest diseases of citrus fruits caused by wound pathogens are green and blue molds and sour rot, caused by the fungi *Penicillium digitatum* (Pers.:Fr.) Sacc., *Penicillium italicum* Wehmer, and *Geotrichum citri-aurantii* (Ferraris) Butler, respectively. Citrus anthracnose, caused by *Colletotrichum gloeosporioides* (Penz.) Penz. & Sacc.; black rot, caused by *Alternaria alternata* (Fr.) Keissl.; and stem-end rots, caused by *Lasiodiplodia theobromae* (Pat.) Griffon and Maubl. and *Phomopsis citri* H.S. Fawc. are important citrus postharvest diseases caused by latent pathogens (Palou 2014; Smilanick et al. 2020). In the case of pome fruits such as apples and pears, blue and gray molds, caused respectively by *Penicillium expansum* L. and *Botrytis cinerea* Pers.:Fr., are among the most deleterious postharvest diseases (Spadaro et al. 2020). *B. cinerea* is also the main causal agent of postharvest decay of grapes, strawberries, blueberries and other small berries, kiwifruits, pomegranates, figs, and some fruit vegetables such as tomatoes, peppers, and eggplants (Palou et al. 2020; Romanazzi et al. 2020; Tzortzakis et al. 2020). *Monilinia* ssp., especially *M. fructicola* (Winter) Honey, *M. laxa* (Aderhold & Ruhland) Honey, and *M. fructigena* (Aderhold & Ruhland) Honey, cause postharvest brown rot of stone fruits such as peaches, nectarines, plums, and sweet cherries (Mari et al. 2020). Many subtropical and tropical fruits, such as avocados, mangoes, papayas, pineapples, and bananas, and many common fruit vegetables are frequently attacked by *Colletotrichum* spp. causing anthracnose, *Alternaria* spp. causing black spot, and *L. theobromae* and other *Botryosphaeriaceae* pathogens causing stem-end rots (Alvindia 2020; Bautista-Baños et al. 2020; Sriram and Sudhakar Rao 2020; Tzortzakis et al. 2020). *A. alternata* is also an important postharvest pathogen of pome fruits, persimmons, loquats, pomegranates, and many other fruits (Troncoso-Rojas and Tiznado-Hernández 2014; Palou et al. 2020; Spadaro et al. 2020). *Rhizopus stolonifer* (Ehrenb.) Vuill., another dangerous postharvest pathogen, is the cause of soft rot on a very wide range of hosts (Bautista-Baños et al. 2014; Tzortzakis et al. 2020).

In general, economic losses due to fungal postharvest diseases have been reduced to commercially acceptable levels by the use of synthetic fungicides such as imazalil (IMZ), thiabendazole (TBZ), pyrimethanil (PYR), fludioxonil (FLU), and other active ingredients. Nevertheless, the continuous use of these agrochemicals is increasingly leading to substantial problems and reduced market opportunities.

Human health issues and environmental pollution due to chemical residues and the proliferation of fungal biotypes resistant to synthetic fungicides are important concerns limiting their use in commercial packinghouses. Therefore, consumer trends and legislation changes are currently favoring a continuous reduction in the amount of these substances allowed by authorities, which makes research necessary for implementation of alternative approaches. In a context where conventional fungicides are not available, effective control will need to adopt integrated strategies that take into account all factors affecting disease epidemiology and incidence, including preharvest factors, in order to minimize decay losses (Palou et al. 2008; Wisniewski et al. 2016). Such a "non-polluting integrated disease management" (NPIDM) concept will be based on the adoption of cost-effective actions in the field, during harvest, and after harvest and should not be confused with the traditional "integrated disease management" (IDM) in the context of agricultural "integrated production," which often implies production in compliance with particular national or regional regulations and programs that still include the use of postharvest conventional fungicides (Palou 2018). Within NPIDM strategies, the relative weight of field actions will be more important in the case of diseases caused by latent pathogens, and actions at harvest will be especially important in the case of diseases caused by wound pathogens, but the commercial adoption of suitable non-polluting postharvest antifungal treatments to replace the use of conventional fungicides will be key in all cases. In general, according to their nature, these alternative treatments can be physical, chemical, or biological (Wisniewski et al. 2016).

Chemical alternatives should be compounds with known and minimal toxicological effects on mammals and impact on the environment and, as substances that will be in contact with fresh produce, these alternative chemicals should be affirmed as generally recognized as safe (GRAS) by the United States Food and Drug Administration (US FDA), as food additives by the European Food Safety Authority (EFSA), or as an equivalent status by national legislations. GRAS materials are exempt from residue tolerances on all agricultural commodities by the US FDA. Low-toxicity chemicals that are recognized as GRAS include some essential oils, plant extracts and other natural compounds, but also synthetic inorganic or organic salts such as carbonates, sorbates, benzoates, acetates, paraben salts, silicates, etc. (Palou et al. 2016a). Although the acidic forms of these compounds may also possess antimicrobial activity, salts are preferred for postharvest treatments because of their superior water solubility, ease of application, and additional activity of cations such as Na^+, K^+, or NH_4^+ (Smilanick et al. 1999). In this chapter, we describe the use for research purposes of this type of salts as alternative low-toxicity means for postharvest preservation of fresh horticultural products, including the assessment of *in vitro* and *in vivo* activity of salt aqueous solutions and the performance as ingredients of edible coatings.

Performance of Salt Aqueous Solutions

The selection of the most convenient GRAS salts for a particular application implies a sequential evaluation research procedure. Different concentrations of salt aqueous solutions can be evaluated *in vitro* to assess their toxicity to mycelium and/or spores of the target postharvest pathogen. Selected salts and concentrations can then be used in *in vivo* laboratory tests with fruits or vegetables artificially inoculated with the pathogen to establish their potential commercial efficacy and the best potential treatment conditions, including synergy with heat (use of heated aqueous solutions). Afterwards, potential commercial implementation of selected postharvest treatments can be assessed in semi-commercial or commercial trials with naturally infected produce.

In Vitro Antimicrobial Activity

Target postharvest pathogens should be isolated, purified, properly identified, and multiplied by replating on appropriate agar culture medium, usually potato dextrose agar (PDA) in the case of fungal pathogens. At the IVIA CTP we have built and maintained a culture collection of postharvest pathogens autochthonous of the Valencian region (Spain), which include fungi causing important diseases on citrus, stone fruits, table grapes, and minor Mediterranean fruits such as persimmon, pomegranate, loquat, and date palm fruit (Palou et al. 2013, 2015a, 2016b, c). Isolations were generally performed from infected fresh fruits found in commercial packinghouses and preliminary *in vivo* pathogenicity tests were conducted with different isolates to select, based on their aggressiveness and uniform behavior, the fungal strains of each species more adequate to be included in the collection and used in the experiments. Appropriate incubation temperatures for growth of most postharvest pathogens on PDA in petri dishes are 20–25 °C. Depending on the growth rate of the different fungal species, the required incubation time to obtain mature fungal inoculum to be used in the experiments is 1–3 weeks.

The most typical procedure for *in vitro* evaluation of the antifungal activity of GRAS salts is the determination of the inhibition of radial mycelial growth by plating the target pathogen on 90-mm diameter plastic petri dishes with PDA medium amended with the test salt at the desired concentration. Since the solubility in water of most GRAS salts is very high, stock solutions of each salt at high concentration (10% or higher) are prepared and diluted to achieve the range of final concentrations that are frequently tested, from 0.1% or lower to a maximum of 2–3%, depending on the characteristics and legal regulations affecting each salt. Always under strict sterile conditions, these solutions are incorporated into the PDA medium at 45–55 °C and poured into the petri dishes to allow solidification. Known spore concentrations or, more commonly, mycelial plugs (around 5 mm in diameter) of the pathogen growing cultures, produced with a sterilized cork borer, are then inoculated at the center of each dish. PDA plates without salt serve as controls.

Inoculated plates are then incubated in a growth cabinet at 20–25 °C for a period of time dependent on the fungal species, but at least until fungal growth completely covers the control plates. Radial mycelial growth is periodically assessed (every 1, 2, or 3 days) in each plate by measuring two perpendicular fungal colony diameters during the entire incubation period. Usually, 3–5 replicate plates are used for each salt and salt concentration. The results are expressed as percentage of mycelial growth inhibition according to the formula: (dc-dt)/dc × 100, where dc = average diameter of the fungal colony on control plates and dt = average diameter of the fungal colony on salt-amended plates. The obtained data are typically subjected to a two-way analysis of variance (ANOVA) with salt and salt concentration as factors. Factor interactions are studied and if they are significant, individual one-way ANOVAs are further performed for the different levels of each factor (Fagundes et al. 2013). With particular salts selected for their high antifungal activity, it can be useful to establish the salt minimum inhibitory concentration (MIC), which implies testing a larger number of concentrations.

In vitro spore mortality or spore germination tests are often used to determine the ability of GRAS salts to kill or inactivate pathogenic fungal spores. In this case, liquid culture medium (PDA broth or similar), amended with GRAS salts at different concentrations and containing a spore suspension of known density (usually 10^4–10^6 spores/mL), is incubated at 20–25 °C for 18–24 h, at which time acid fuchsin solution is added to stop further germination. Germinated spores (those with the germ tube length equal or longer than the spore itself) are then counted in an inverted compound microscope. Control treatment consists of medium broth without GRAS salt. Data are generally expressed as percent spore germination inhibition and calculated as previously described for the percent mycelial growth inhibition. Each concentration of each salt is commonly applied to 3–5 replicates and microwell plates are often used to test many salt concentrations, which will allow the determination of the effective doses that kill 50 or 95% (ED_{50}, ED_{95}) of the spores (Smilanick et al. 1999; Youssef and Roberto 2014).

Results of *in vitro* evaluations are taken into account to select the most suitable GRAS salts and concentrations to be tested as aqueous solutions in *in vivo* experiments with fruit. Our research group at the IVIA CTP has worked in collaboration with other groups from Brazil, Mexico, and Turkey on the *in vitro* selection of GRAS salts with activity against *B. cinerea* (Table 11.1, Fagundes et al. 2013), *A. alternata* (Fagundes et al. 2013), *M. fructicola* (Karaca et al. 2014), or *L. theobromae* (Guimarães et al. 2019).

Table 11.1 *In vitro* antifungal activity of GRAS salts amended at different concentrations to PDA plates against *Botrytis cinerea* after 5 days of incubation at 25 °C

	Inhibition of *B. cinerea* (%)[a]		
	Salt concentration (%)		
GRAS salt	0.2	1.0	2.0
Sodium formate	0.00 cE	28.27 bC	55.26 aC
Potassium silicate	94.32 aA	94.32 aA	94.32 aA
Sodium acetate	7.81 cD	39.63 bB	67.90 aB
Sodium propionate	57.39 bB	87.07 aA	94.18 aA
Sodium methylparaben	94.32 aA	94.32 aA	94.32 aA
Sodium ethylparaben	94.32 aA	94.32 aA	94.32 aA
Sodium propylparaben	94.32 aA	94.32 aA	94.32 aA
Potassium sorbate	58.52 bB	94.32 aA	94.32 aA
Sodium benzoate	39.20 bC	94.32 aA	94.32 aA

Means in lines with different lowercase letters and means in columns with different capital letters are significantly different by Fisher's protected LSD test ($P < 0.05$) applied after an ANOVA. (Reproduced with permission from Elsevier from Fagundes et al. 2013)
[a]Colony diameter reduction with respect to control treatments (non-amended PDA plates)

In Vivo Postharvest Disease Control

Postharvest treatments using aqueous solutions of GRAS salts with antifungal properties are interesting because the replacement of conventional chemical fungicides with these products would not require substantial changes in the industrial procedures followed in fresh produce packinghouses (Palou et al. 2016a). Different types of *in vivo* tests can be designed depending on the objective and scale of each particular investigation, but in all cases this research involves, by definition, the use of harvested fruits and vegetables. These are usually collected from commercial orchards and used the same or the following days. It is important, before each experiment, to properly select, randomize, wash, superficially disinfect, and rinse the produce samples.

In general, two different kinds of antifungal activity can be assessed in *in vivo* tests: curative activity, for which fruit artificially inoculated with the pathogen are treated with the salt solution after different periods of time; and preventive or protectant activity, for which fruit treated with the salt solution are artificially inoculated with the pathogen after different periods of time (Moscoso-Ramírez and Palou 2014). A period of 24 h is frequently used for diseases caused by wound pathogens since it resembles well the time between potential infections in the field, commonly at harvest time, and potential treatments in the packinghouse (Eckert and Brown 1986).

Generally, it can be useful to start the *in vivo* research with primary screenings; i.e., small-scale laboratory tests designed to select the best GRAS salt and salt concentration for a particular type of fruit species or cultivar. In some cases, *in vitro* tests can be replaced with this type of *in vivo* primary screenings. A highly-concentrated sterile mother solution of the salt is prepared to obtain by dilution the desired concentrations to be tested. The salt solution is applied with a micropipette in wounds inflicted in the rind of mature fruit after (for assessment of curative activity) or before (for assessment of preventive activity) the inoculation of the target pathogen in these rind wounds. Typically, fungal inoculation is performed by applying into the rind wound a small volume (10–30 μL) of a conidial suspension of known concentration (10^4–10^6 spores/mL). Control fruit are inoculated and treated with sterile distilled water. Depending on the fruit size and shape, one or more rind wounds per fruit can be inflicted. Usual sample size for this type of tests is of 2–4 replicates of 5–10 fruit each. Treated fruit are incubated at 20–25 °C and high relative humidity (RH) for a variable period of time (until most of control fruit are actually rotten), which can be approximately of 1–3 weeks at these incubation conditions (Moscoso-Ramírez et al. 2013). Disease incidence (percentage of infected wounds) and severity (lesion diameter) and pathogen sporulation (percentage of lesions showing spores) are determined every 3–7 days. Results can be also expressed as percent reductions with respect to control fruit, especially if results of several independent experiments are compiled (Eckert and Brown 1986).

Small-scale trials are typically conducted with fresh fruit artificially inoculated with the target pathogen to establish the best postharvest treatment conditions to resemble potential commercial applications in fresh produce packinghouses. In general, dip or flood treatments are evaluated for research purposes with aqueous solutions since they are the most effective application method, but in some cases other application systems such as drencher, spray, and low pressure aqueous curtain are tested. Depending on the experiment, inoculated and treated fruit can be incubated at 20–25 °C to favor fast fungal development or long-term stored at low temperatures to resemble commercial fruit handling. Again, both curative and preventive activities can be tested in this type of experiments. Since GRAS salts aqueous solutions are often synergic with heat for disease control, dip treatments of different length (from 30 s to 3 min) with solutions heated at different temperatures (45–60 °C) can be tested (Palou et al. 2001, 2002; Montesinos-Herrero et al. 2009). For this, in the IVIA CTP, a stainless steel water tank, fitted with electrical resistances and a thermostat, able to heat fruit placed in steel buckets is used. Control fruit are typically treated with water at 20 °C for 30–60 s. Usual sample size for these trials is of 3–5 replicates of 20–25 fruit each. Treated fruit are commonly arranged in plastic cavity sockets on cardboard or plastic trays before incubation or cold storage. Disease incidence and severity and pathogen sporulation are periodically determined during these periods.

As a final conclusive step, semi-commercial or commercial trials can be conducted with naturally infected fruit and larger sample sizes than in previous trials. Research pilot plants with equipment, such as fruit packinglines or drenchers, which resemble at small scale those in commercial packinghouses, are used for semi-commercial trials, with a sample size dependent on the type of produce; e.g., 3–5 replicates of 100–300 fruit per treatment. Commercial trials are directly conducted in industrial facilities with several replications per treatment of various entire fruit field boxes or packages. In every case, treated fruit are commercially handled in the packinghouse and follow the same postharvest steps than fruit used for actual commercialization (grading, cold storage, etc.). The performance of GRAS treatments is determined in terms of reduction of total decay, reduction of particular diseases, and effects on the quality of commercial produce.

Depending on the nature of the response or dependent variables, data from the above trials are typically analyzed by multifactor ANOVA and means separation tests or generalized linear models (GLM) such as binary logistic regression. Disease incidence and severity and pathogen sporulation are used as dependent variables to establish the effectiveness of GRAS salts. Sometimes these variables are expressed as percentages of reduction with respect to the control fruit. This can allow joining results from different analog trials. Depending on the experiment, additional variables that are taken into account are incidence and severity of potential phytotoxicity induced to the fruit peel by the treatment, and fruit quality attributes during the cold storage period and after shelf life.

For many years, our research group at the IVIA CTP has worked in the evaluation of aqueous solutions of different GRAS salts, especially against citrus green and blue molds caused by *Penicillium* spp. Thus, the effectiveness of solutions of carbonates and bicarbonates (Palou et al. 2001, 2002), sorbates (Fig. 11.1, Montesinos-Herrero et al. 2009), benzoates (Montesinos-Herrero et al. 2016), propionates (Palou et al. 2002), silicates (Moscoso-Ramírez and Palou 2014), or sodium parabens (Moscoso-Ramírez et al. 2013) at room temperature or heated to non-phytotoxic temperatures has been determined. A large variety of GRAS salts were also evaluated against the most important postharvest fungal diseases of stone fruit (Palou et al. 2009).

GRAS Salts as Ingredients of Edible Coatings

GRAS salts are increasingly gaining importance as antifungal ingredients for the development of novel synthetic antimicrobial edible coatings based on natural biopolymers and lipids for fresh horticultural products. These coatings provide the dual action of creating a partial barrier to water vapor and gases such as oxygen and carbon dioxide and preventing or reducing microbial growth during storage, which translates into increased postharvest life (Valencia-Chamorro et al. 2011; Pérez-Gago and Palou 2016).

In general, basic components of edible coatings for fruits and vegetables are blends of hydrocolloids (proteins or polysaccharides), lipids (waxes, acylglycerols, or fatty acids), and occasionally some resins as shellac intended to provide gloss (Valencia-Chamorro et al. 2011). Beside these main ingredients constituting the composite coating matrix, plasticizers (e.g., sucrose, glycerol, sorbitol, propylene glycol, polyethylene glycol, or monoglycerides) and emulsifiers (e.g., fatty acids, ethylene glycol monostearate, glycerol monostearate, esters of fatty acids, lecithin, sucrose ester, and sorbitan monostearate or polysorbates) are key minor ingredients that improve coating integrity and emulsion stability, respectively (Han 2014; Hassan et al. 2018). Additional ingredients can be added to provide other functional

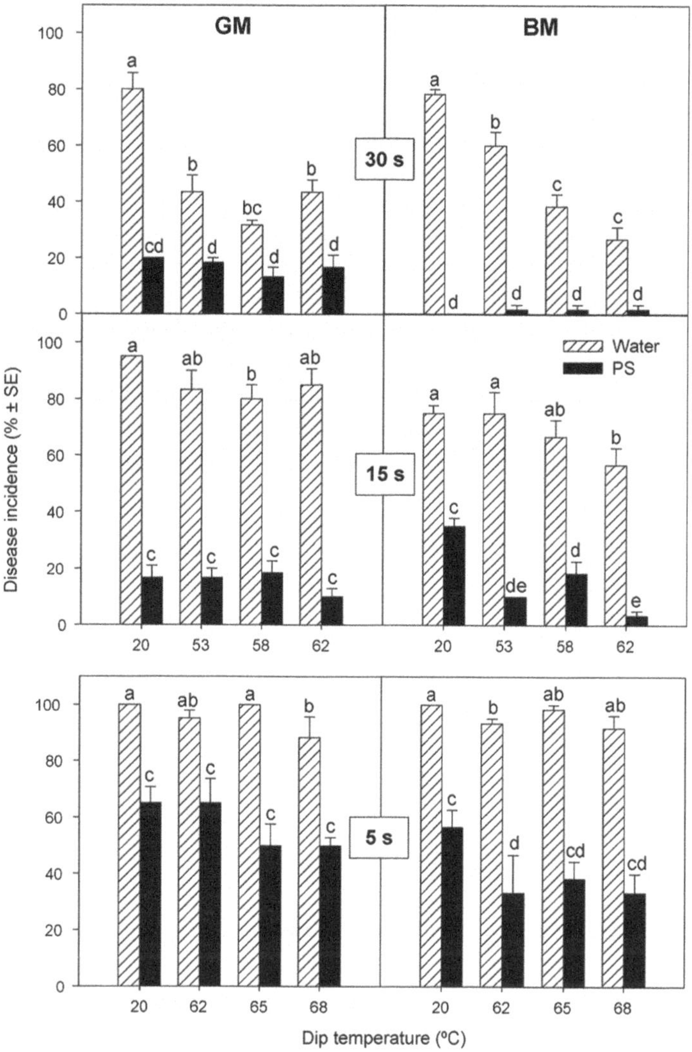

Fig. 11.1 Incidence of green (GM) and blue (BM) molds on 'Valencia' oranges artificially inoculated with *Penicillium digitatum* and *Penicillium italicum*, respectively, dipped 24 h later in water or aqueous solutions of 3% (w/v) potassium sorbate (PS) at different temperatures for 5, 15, or 30 s, and incubated for 7 days at 20 °C and 90% RH. For each disease and treatment time, columns with unlike letters are significantly different according to Fisher's protected LSD test ($P < 0.05$) applied to arcsine-transformed data. Non-transformed means±SE are shown. (Reproduced with permission from Elsevier from Montesinos-Herrero et al. 2009)

properties to the coating, such as antioxidants, nutrients, nutraceuticals, and antimicrobials. The latter include natural compounds such as essential oils, spices, and a large variety of plant extracts; antimicrobial peptides and small proteins; some nanostructures including metal-based nanoparticles (Ag, ZnO, TiO_2, SiO_2, etc.); and synthetic food additives or GRAS compounds (Palou et al. 2016a; González-Reza et al. 2018; Dukare et al. 2019).

The evaluation of particular antimicrobial edible coatings should consider the interactions among all the elements of the system, which include the target pathogen, the coating ingredients, the antimicrobial agent, and the characteristics and usual postharvest handling of the horticultural product. Therefore, the development of these coatings requires an initial optimization of coating formulations based on chemical compatibility of the ingredients to achieve stable emulsions. The physiological postharvest characteristics of the particular fresh product to be coated will determine the non-antifungal ingredients of the coating, while the GRAS salts and concentrations will be selected depending on previous information about the control ability of salt aqueous solutions for that particular pathosystem. Coating formulations are usually optimized on the basis of percent total solid content, total lipid content, and GRAS salt concentration, but other parameters such as viscosity, pH, and wettability for the particular commodity are also important (Valencia-Chamorro et al. 2008, 2011). Compatible and stable formulations can then be used to form films for *in vitro* studies or can be directly applied to the commodity for *in vivo* studies. Incompatible emulsions solidify or show phase separation or undesirable physical characteristics. Thus, for example, Valencia-Chamorro et al. (2008) reported that only around 5% of 470 edible coating formulations based on hydroxypropyl methylcellulose (HPMC)-lipid formed stable emulsions with appropriated viscosity for citrus coating application when GRAS salts, selected from previous studies with aqueous solutions, were incorporated as antifungal ingredients. Such optimization was based on the lipid concentration (in a range of 0–60%, dry basis), the emulsion solid concentration (in a rate of 4–12%), and GRAS salt concentration (0.05–4.5%, wet basis).

In Vitro Antimicrobial Activity of Films

Although films and coatings have the same chemical composition and sometimes are used as synonymous, they refer to different concepts according to their different purpose and utilization. Films are defined as a stand-alone thin layer of materials prepared by casting process and used as covers, wraps, or separation layers. They are usually used for determination of barrier, mechanical, and other properties of the coating formulation. On the other hand, fruit coatings involve the formation of films directly on the surface of the fruit to which they are intended to be applied (Palou et al. 2015b). Therefore, films are appropriate for the *in vitro* determination of the antimicrobial activity of edible coatings, which is usually tested by means of the disk diameter test.

For determination of antifungal activity, coating emulsions are casted onto plastic plates to dry in sterile conditions. Dry films are peeled intact from the casting surface and aseptically cut into disks (5–15 mm diameter) using a sterile cork borer. Films from the emulsion matrix without the antifungal GRAS salt are used as controls. Film disks are then aseptically transferred to the surface of agar culture medium petri dishes, typically PDA, previously inoculated with spores of the target pathogen. The dishes are refrigerated at 4–5 °C for few hours to allow for the diffusion of film ingredients and then incubated at 20–25 °C for 1–2 weeks. Usually, 3–4 agar plates (replicates) are prepared for each pathogen and film. The antifungal activity of the film is determined by periodically measuring the length of the inhibition zone around the film disk (Valencia-Chamorro et al. 2008). Results from disk diameter tests allow comparing in a rapid and simple way different salts and salt concentrations in the emulsion matrix or different matrixes with the same antifungal salt.

In Vivo Performance of Coatings

When the number of coatings or coating antimicrobial ingredients to consider for a particular application is not very high, it can be more practical to evaluate the control ability of the coatings directly by *in vivo* tests. Although more laborious, time-consuming, and expensive, *in vivo* screenings will be always more accurate than disk diameter tests because they simulate much better than pathogens growing in artificial media the actual conditions and complex interactions among host (fresh horticultural product), pathogen, and environment that occur during disease development in infected produce. Furthermore, coating performance is not only dependent on intrinsic characteristics of the formulated emulsion, but also on characteristics of the host, especially surface properties of the cuticle and whole peel of fruits and vegetables. In general, these properties can greatly influence the diffusion rate of GRAS salts or other antimicrobial agents to the infection courts occupied by the pathogen in the host peel. Thus, the release capability of antimicrobial agents from films located on agar medium can differ from that of coatings located on the fresh product peel surface, and the overall performance of coating formulations is typically highly dependent on the commodity species and even cultivar (Pérez-Gago and Palou 2016).

As an example of the research sequence, Figure 11.2 shows a schematic diagram for the development/selection of HPMC-lipid based antifungal edible coatings incorporating GRAS salts as antifungal agents for the control of postharvest diseases and quality maintenance of fresh fruits during long-term cold storage. Once coating emulsions containing GRAS salts are formulated and proved stable, their disease control ability is evaluated in *in vivo* antifungal tests with fruit artificially inoculated with the target pathogen and incubated at 20–25 °C. The most effective coatings are selected and both physico-chemical and sensory fruit quality attributes of cold-stored fruit are then evaluated to select the most appropriate antifungal edible coating for overall postharvest preservation of the target fresh commodity.

General procedures for *in vivo* disease control assessment are as follows: samples of the target fresh product are selected, washed, artificially inoculated with the target pathogen, and, after about 24 h (curative activity), coated with the different

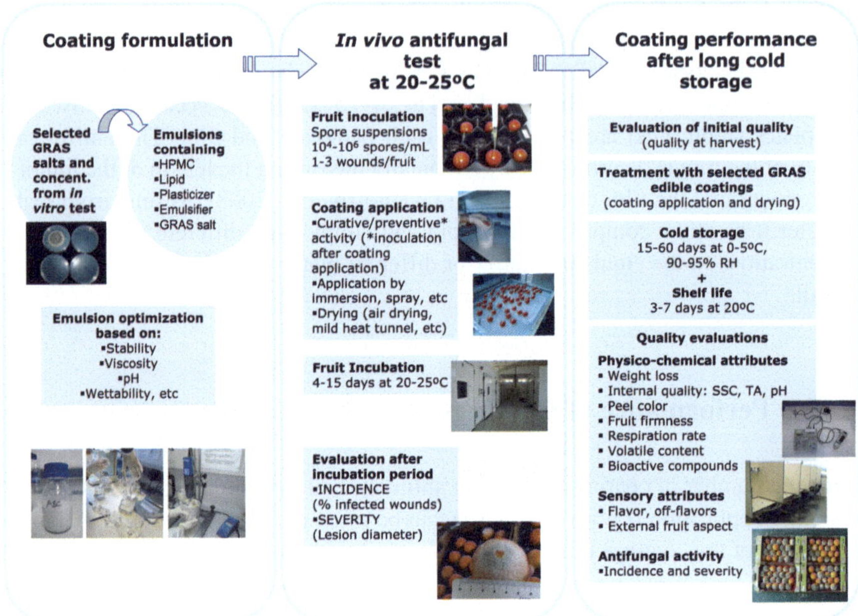

Fig. 11.2 Diagram of the development of hydroxypropyl methylcellulose (HPMC)-lipid based edible coatings incorporating GRAS salts as antifungal agents for fresh fruits

coating treatments, and allowed to dry on a mesh screen (air drying) or dried in a mild heat tunnel, commonly at 35–45 °C. Although considerably less usual, preventive activity could also be evaluated by inoculating the pathogen for a variable period of time after applying the coating. Coatings are usually applied by dipping the product in the emulsion for brief periods (10–30 s) (Valencia-Chamorro et al. 2009), but they can also be applied by pipetting a small amount of the emulsion (0.1–0.5 mL) onto each fruit and rubbing manually with gloved hands to mimic spray coating application in industrial packingline roller conveyors (Gunaydin et al. 2017). Control fruit are inoculated but, depending on the experiment and its particular objectives, uncoated (or treated with water) or treated with coatings formulated without GRAS salts. Typically, inoculated and treated fruit are incubated at 20–25 °C (optimal temperatures for fungal growth and consequent postharvest disease development; Eckert and Brown 1986, Palou 2018), which allow to obtain results in about 1–2 weeks. Sample size, inoculation, incubation, and evaluation procedures, dependent variables, and statistical analyses used for these *in vivo* trials are equivalent to those described above for the *in vivo* evaluation of disease control ability of GRAS salt aqueous solutions.

Once the most effective coatings to control the target disease are identified, their effect on the physiological behavior and quality of treated and cold-stored produce needs to be determined to select the most appropriate antifungal coatings for overall produce preservation. After determination of the initial quality (quality at harvest), fresh produce is coated and subjected to cold storage (generally 15–60 days at 0–5 °C and 90–95% RH, depending on the commodity) and simulated periods of shelf life (3–7 days at 20 °C). Both physicochemical and sensory quality attributes are periodically evaluated during storage and after shelf life. The most common physicochemical quality attributes include weight loss; peel color; firmness; juice internal quality parameters such as soluble solids content (SSC), titratable acidity (TA), and pH; respiration rate; internal gas concentration (O_2 and CO_2); volatile content (ethanol and acetaldehyde); and in some cases relevant bioactive compounds (nutrients, vitamins, antioxidants, etc.). Sensory fruit attributes can be determined by trained expert panels or consumer tests and include flavor, off-flavors, and external and internal visual aspect. In some cases, the assessment of disease parameters such as incidence, severity, and pathogen sporulation on artificially inoculated or naturally infected and cold-stored produce can also be of interest.

Research by our group at the IVIA CTP showed that HPMC-lipid composite coatings containing GRAS salts were effective for controlling citrus green and blue molds caused by *P. digitatum* and *P. italicum*, respectively, on commercial orange and mandarin cultivars (Valencia-Chamorro et al. 2008, 2009). In a subsequent stage, the physiological performance of selected coatings was tested during citrus long-term cold storage (30–60 days at 5 °C plus 7 days of shelf life simulation at 20 °C) (Valencia-Chamorro et al. 2010). Overall, the performance of the coatings strongly depended on the citrus species and cultivar. In general, the most effective coatings to inhibit Penicillium molds were those containing the salts potassium sorbate, sodium benzoate, sodium propionate, or mixtures of them. Likewise, the addition of the GRAS salts affected differently the gas and moisture barrier of the coatings and the overall quality of coated fruit during cold storage. In more recent work, similar coatings formulated with potassium sorbate or sodium benzoate also resulted effective to reduce citrus stem-end rot caused by *L. theobromae* (Fig. 11.3, Guimarães et al. 2019). Similar studies proved the antifungal activity of several GRAS salts incorporated to HPMC-beeswax based coatings against *B. cinerea* and *A. alternata* on artificially inoculated cherry tomato fruit during incubation at 20–25 °C (Fagundes et al. 2013). These antifungal HPMC-based coatings also reduced gray mold and black spot development on inoculated cherry tomatoes stored at 5 °C for 15 days, with coatings containing the salts sodium propylparaben and sodium benzoate, respectively, the most effective. However, only coatings formulated with ammonium carbonate and sodium benzoate were effective to control weight loss and maintain the firmness of coated cherry tomatoes (Fagundes et al. 2014, 2015). Among a wide variety of antifungal GRAS salts tested as ingredients of HPMC-beeswax edible coatings, paraben salts and potassium sorbate were the best agents for effective control of brown rot caused by *M. fructicola* on artificially inoculated plums (Karaca et al. 2014). These coatings also effectively reduced plum weight and firmness loss and, after 22 days of storage at 1 °C followed by a shelf life period of 5 days at 20 °C, coated fruit showed higher TA, SSC, and hue angle values, as well as reduced physiological disorders, than uncoated control fruit (Gunaydin et al. 2017).

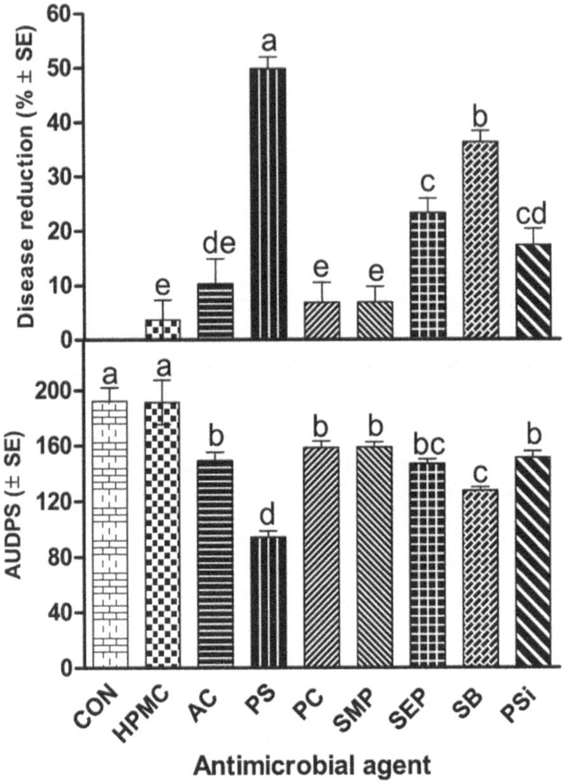

Fig. 11.3 Reduction of disease severity with respect to control fruit (inoculated but uncoated) and area under the disease progress stairs (AUDPS) of stem-end rot on 'Ortanique' mandarins artificially inoculated with *Lasiodiplodia theobromae*, coated 24 h later with HPMC-beeswax composite edible coatings containing GRAS salts, and incubated for 10 days at 28 °C and 90% RH. *AC* ammonium carbonate, *PS* potassium sorbate, *PC* potassium carbonate, *SMP* sodium methylparaben, *SEP* sodium ethylparaben, *SB* sodium benzoate, *PSi* potassium silicate. AUDPS was determined by measuring of lesion diameter after 3, 7, and 10 days of incubation. Columns with different letters are significantly different according to Fisher's protected LSD test ($P < 0.05$) applied after an ANOVA. (Reproduced with permission from Elsevier from Guimarães et al. 2019)

Acknowledgements Spanish IVIA, INIA, and AEI and the European Union Commission (FEDER Program) are gratefully acknowledged for providing financial support to conduct research on this topic.

References

Alvindia DG (2020) Banana and plantain. In: Palou L, Smilanick JL (eds) Postharvest pathology of fresh horticultural produce. CRC Press, Taylor and Francis Group, Boca Raton, pp 277–301

Bautista-Baños S, Bosquez-Molina E, Barrera-Necha LL (2014) *Rhizopus stolonifer* (soft rot). In: Bautista-Baños S (ed) Postharvest decay, Control strategies. Academic Press, Elsevier, London, pp 1–44

Bautista-Baños S, Ventura-Aguilar RI, Ramos-García ML (2020) Avocado. In: Palou L, Smilanick JL (eds) Postharvest pathology of fresh horticultural produce. CRC Press, Taylor and Francis Group, Boca Raton, pp 227–256

Dukare A. S, Paul S, Nambi V. E et al (2019) Exploitation of microbial antagonists for the control of postharvest diseases of fruits: a review. Crit Rev Food Sci Nutr 59:1498–1513

Eckert JW, Brown GE (1986) Evaluation of postharvest fungicide treatments for citrus fruits. In: Hickey KD (ed) Methods for evaluating pesticides for control of plant pathogens. APS Press, St. Paul, pp 92–97

Fagundes C, Pérez-Gago MB, Monteiro AR et al (2013) Antifungal activity of food additives in vitro and as ingredients of hydroxypropyl methylcellulose-lipid edible coatings against *Botrytis cinerea* and *Alternaria alternata* on cherry tomato fruit. Int J Food Microbiol 166:391–398

Fagundes C, Palou L, Monteiro AR et al (2014) Effect of antifungal hydroxypropyl methylcellulose-beeswax edible coatings on gray mold development and quality attributes of cold-stored cherry tomato fruit. Postharvest Biol Technol 92:1–8

Fagundes C, Palou L, Monteiro AR et al (2015) Hydroxypropyl methylcellulose-beeswax edible coatings formulated with antifungal food additives to reduce alternaria black spot and maintain postharvest quality of cold-stored cherry tomatoes. Sci Hortic 193:249–257

González-Reza RM, García-Betanzos CI, Sánchez-Valdés LI et al (2018) The functionalization of nanostructures and their potential applications in edible coatings. Coatings 8:160

Guimarães JER, de la Fuente B, Pérez-Gago MB et al (2019) Antifungal activity of GRAS salts against *Lasiodiplodia theobromae* in vitro and as ingredients of hydroxypropyl methylcellulose-lipid composite edible coatings to control Diplodia stem-end rot and maintain postharvest quality of citrus fruit. Int J Food Microbiol 301:9–18

Gunaydin S, Karaca H, Palou L et al (2017) Effect of hydroxypropyl methylcellulose-beeswax composite edible coatings formulated with or without antifungal agents on physicochemical properties of plums during cold storage J Food Qual: Article ID 8573549

Han JH (2014) Edible films and coatings. In: Han JH (ed) Innovations in food packaging. Academic Press, Elsevier, Amsterdam, pp 234–255

Hassan B, Chatha SAS, Hussain AI et al (2018) Recent advances on polysaccharides, lipids and protein based edible films and coatings: a review. Int J Biol Macromol 109:1095–1107

Karaca H, Pérez-Gago MB, Taberner V et al (2014) Evaluating food additives as antifungal agents against *Monilinia fructicola* in vitro and in hydroxypropyl methylcellulose-lipid composite edible coatings for plums. Int J Food Microbiol 179:72–79

Mari M, Spadaro D, Casals C et al (2020) Stone fruits. In: Palou L, Smilanick JL (eds) Postharvest pathology of fresh horticultural produce. CRC Press, Taylor and Francis Group, Boca Raton, pp 111–140

Montesinos-Herrero C, del Río MA, Pastor C et al (2009) Evaluation of brief potassium sorbate dips to control postharvest penicillium decay on major citrus species and cultivars. Postharvest Biol Technol 52:117–125

Montesinos-Herrero C, Moscoso-Ramírez PA, Palou L (2016) Evaluation of sodium benzoate and other food additives for the control of citrus postharvest green and blue molds. Postharvest Biol Technol 115:72–80

Moscoso-Ramírez PA, Palou L (2014) Preventive and curative activity of postharvest potassium silicate treatments to control green and blue molds on orange fruit. Eur J Plant Pathol 138:721–732

Moscoso-Ramírez PA, Montesinos-Herrero C, Palou L (2013) Characterization of postharvest treatments with sodium methylparaben to control citrus green and blue molds. Postharvest Biol Technol 77:128–137

Palou L (2014) *Penicillium digitatum*, *Penicillium italicum* (green mold, blue mold). In: Bautista-Baños S (ed) Postharvest decay, Control strategies. Academic Press, Elsevier, London, pp 45–102

Palou L (2018) Postharvest treatments with GRAS salts to control fresh fruit decay. Horticulturae 4:46

Palou L, Smilanick JL, Usall J et al (2001) Control of postharvest blue and green molds of oranges by hot water, sodium carbonate, and sodium bicarbonate. Plant Dis 85:371–376

Palou L, Usall J, Smilanick JL et al (2002) Evaluation of food additives and low-toxicity compounds as alternative chemicals for the control of *Penicillium digitatum* and *Penicillium italicum* on citrus fruit. Pest Manag Sci 58:459–466

Palou L, Smilanick JL, Droby S (2008) Alternatives to conventional fungicides for the control of citrus postharvest green and blue moulds. Stewart Postharv Rev 4:1–16

Palou L, Smilanick JL, Crisosto CH (2009) Evaluation of food additives as alternative or complementary chemicals to conventional fungicides for the control of major postharvest diseases of stone fruit. J Food Prot 72:1037–1046

Palou L, Taberner V, Guardado A et al (2013) Incidence and etiology of postharvest fungal diseases of pomegranate (*Punica granatum* cv. Mollar de Elche) in Spain. Phytopathol Mediterr 52:478–489

Palou L, Montesinos-Herrero C, Tarazona I et al (2015a) Incidence and etiology of postharvest fungal diseases of persimmon (*Diospyros kaki* Thunb. cv. Rojo brillante) in Spain. Plant Dis 99:1416–1425

Palou L, Valencia-Chamorro SA, Pérez-Gago MB (2015b) Antifungal edible coatings for fresh citrus fruit: a review. Coatings 5:962–986

Palou L, Ali A, Fallik E et al (2016a) GRAS plant- and animal-derived compounds as alternatives to conventional fungicides for the control of postharvest diseases of fresh horticultural produce. Postharvest Biol Technol 122:41–52

Palou L, Rosales R, Taberner V et al (2016b) Incidence and etiology of postharvest diseases of fresh fruit of palm (*Phoenix dactylifera* L.) in the grove of Elx (Spain). Phytopathol Mediterr 55:391–400

Palou L, Sánchez-Torres P, Montesinos-Herrero C et al (2016c) Incidence and etiology of postharvest fungal diseases of loquat fruit (*Eriobotrya japonica* (Thunb.) Lindl. cv. 'Algerie') in Alacant province (Spain). Eur J Plant Pathol 146:847–860

Palou L, Kinay-Teksür P, Cao S et al (2020) Pomegranate, persimmon, and loquat. In: Palou L, Smilanick JL (eds) Postharvest pathology of fresh horticultural produce. CRC Press, Taylor and Francis Group, Boca Raton, pp 187–225

Pérez-Gago MB, Palou L (2016) Antimicrobial packaging for fresh and fresh-cut fruits and vegetables. In: Pareek S (ed) Fresh-cut fruits and vegetables: technology, physiology, and safety. CRC Press, Taylor and Francis Group, Boca Raton, pp 403–452

Romanazzi G, Elmer PAG, Feliziani E (2020) Table grape, kiwifruit, and strawberry. In: Palou L, Smilanick JL (eds) Postharvest pathology of fresh horticultural produce. CRC Press, Taylor and Francis Group, Boca Raton, pp 141–185

Smilanick JL, Margosan DA, Mlikota-Gabler F et al (1999) Control of citrus green mold by carbonate and bicarbonate salts and the influence of commercial postharvest practices on their efficacy. Plant Dis 83:139–145

Smilanick JL, Erasmus E, Palou L (2020) Citrus fruits. In: Palou L, Smilanick JL (eds) Postharvest pathology of fresh horticultural produce. CRC Press, Taylor and Francis Group, Boca Raton, pp 3–53

Spadaro D, Torres R, Errampalli D et al (2020) Pome fruits. In: Palou L, Smilanick JL (eds) Postharvest pathology of fresh horticultural produce. CRC Press, Taylor and Francis Group, Boca Raton, pp 55–109

Sriram S, Sudhakar Rao DV (2020) Papaya. In: Palou L, Smilanick JL (eds) Postharvest pathology of fresh horticultural produce. CRC Press, Taylor and Francis Group, Boca Raton, pp 257–276

Troncoso-Rojas R, Tiznado-Hernández ME (2014) *Alternaria alternata* (black rot, black spot). In: Bautista-Baños S (ed) Postharvest decay, Control strategies. Academic Press, Elsevier, London, pp 147–187

Tzortzakis N, Alkan N, Ziv C et al (2020) Solanaceae and Cucurbitaceae crops. In: Palou L, Smilanick JL (eds) Postharvest pathology of fresh horticultural produce. CRC Press, Taylor and Francis Group, Boca Raton, pp 303–337

Valencia-Chamorro SA, Palou L, del Río MA et al (2008) Inhibition of *Penicillium digitatum* and *Penicillium italicum* by hydroxypropyl methylcellulose-lipid edible composite films containing food additives with antifungal properties. J Agric Food Chem 56:11270–11278

Valencia-Chamorro SA, Pérez-Gago MB, del Río MA et al (2009) Curative and preventive activity of hydroxypropyl methylcellulose-lipid edible composite coating antifungal food additives to control citrus postharvest green and blue molds. J Agric Food Chem 57:2770–2777

Valencia-Chamorro SA, Pérez-Gago MB, del Río MA et al (2010) Effect of antifungal hydroxypropyl methylcellulose (HPMC)-lipid edible composite coatings on penicillium decay development and postharvest quality of cold-stored 'Ortanique' mandarins. J Food Sci 75:418–426

Valencia-Chamorro SA, Pérez-Gago MB, del Río MA et al (2011) Antimicrobial edible films and coatings for fresh and minimally processed fruits and vegetables: a review. Crit Rev Food Sci Nutr 51:872–900

Wisniewski M, Droby S, Norelli J et al (2016) Alternative management technologies for postharvest disease control: the journey from simplicity to complexity. Postharvest Biol Technol 122:3–10

Youssef K, Roberto SR (2014) Salt strategies to control *Botrytis* mold of 'Benitaka' table grapes and to maintain fruit quality during storage. Postharvest Biol Technol 95:95–102

Chapter 12
Electrolyzed Water as a Potential Agent for Controlling Postharvest Decay of Fruits and Vegetables

Antonio Ippolito, Annamaria Mincuzzi, Antony Surano, Khamis Youssef, and Simona Marianna Sanzani

Abstract Disinfection after harvest is an essential step to maintain commodities and facilities free of fungal and bacterial postharvest pathogens, responsible of storage decay and economic losses. Electrolyzed water (EW) has gained considerable interest over the last decades as a novel broad-spectrum sanitizer. EW is sustainable and cost effective since it can be produced on-site utilizing tap water and different inexpensive salts and is healthy for both the environment and human beings. Its effectiveness in controlling fungi, yeasts, and bacteria within a wide range of pH is due to multiple mode of actions. Furthermore, its strong oxidizing potential is capable to reduce the amount of pesticide residues on fruit and vegetable surfaces and to avoid pathogen resistance. Properties of EW are related to salts employed for production, being those with low chlorine content preferable. Lastly, EW has no negative effect on the organoleptic properties and features of treated commodities. The present chapter highlights recent developments in EW generation, factors affecting its effectiveness for controlling postharvest decay of fruits and vegetables, mechanism of action on microbes and hosts, and advantages and disadvantages on its use.

Keywords Physical means · Electrolyzed water · Sodium bicarbonate · Sodium chloride · Fruits and vegetables · Decontamination

---

A. Ippolito (✉) · A. Mincuzzi · A. Surano
Dipartimento di Scienze del Suolo, della Pianta e degli Alimenti, Università degli Studi di Bari Aldo Moro, Bari, Italy
e-mail: antonio.ippolito@uniba.it

K. Youssef
Plant Pathology Research Institute, Agricultural Research Center, Giza, Egypt

S. M. Sanzani
CIHEAM-Bari, Valenzano, BA, Italy

D. Spadaro et al. (eds.), *Postharvest Pathology*, Plant Pathology in the 21st Century 11, https://doi.org/10.1007/978-3-030-56530-5_12

Introduction

Postharvest decay of fruits and vegetables is often a direct result of poor handling practices in the packinghouse environment. The wash water used in dump tanks for processing is among the various sources of pathogen contamination; thus, its proper sanitation is extremely important for delivering healthy products to the consumer and minimizing postharvest losses. Indeed, sanitation after harvest can reduce spoilage by 50% or more (Sargent et al. 2000). The most popular disinfecting agent is chlorine (hypochlorite) applied as spray or dip, but several alternative sanitizers of minor use during washing or storage of fresh produce are available, such as chlorine dioxide, ozone, ethanol, hydrogen peroxide, organic acids, and electrolyzed water (EW). This latter has gained importance in the food industry, representing a relevant technical advancement (Buck et al. 2002; Hricova et al. 2008; Feliziani et al. 2016; Rahman et al. 2016). It was firstly developed in Russia for water decontamination and regeneration (Kunina 1967), then it gained great interest for sterilization of utensils, meats, cutting boards, and, more recently, in livestock management and for the sanitation of the washing waters of fresh and minimally processed fruit and vegetables (Lee et al. 2004; Guentzel et al. 2010; Fallanaj et al. 2013; Gómez-López et al. 2013). This chapter will address EW generation, factors affecting its effectiveness, mechanism of action on microbes and hosts, and advantage and disadvantage on its use.

Generation of EW

In chemistry, the electrolysis is the process by which electrical energy is transformed into chemical energy, where an electric current passes through an electrolyte with subsequent migration of positive and negatively charged ions towards the negative and positive electrodes, respectively. EW is typically produced by electrolysis of dilute solutions of sodium chloride (NaCl) in an electrolysis cell with or without a diaphragm, which separates the anode (+) and cathode (−). Salts such as potassium chloride (KCl), magnesium chloride ($MgCl_2$), sodium sulfite (Na_2SO_3), sodium hydrogen carbonate ($NaHCO_3$) and many others (Table 12.1) can also be used (Buck et al. 2002; Fallanaj et al. 2013; Feliziani et al. 2016; Youssef and Hussien 2020). However, since a certain amount of chloride is contained in tap water, it would be possible to reactivate its free chlorine by electrolysis, although obtained amount is usually too low to be effective against pathogenic microorganisms (Nakajima et al. 2004). Recently a PE-1 water ionizer machine (Shenzhen, Guangdong, China) that use only naturally present salts in tap water allowed to obtain good results by selecting different levels of electrolyzing potentials (Hussien et al. 2017).

In an electrolysis cell divided by a membrane, two types of EW are produced: the acidic electrolyzed water (AEW) and the basic electrolyzed water (BEW), as displayed in Fig. 12.1. During electrolysis, the dissociated Cl^- together with OH^- move

Table 12.1 Salts utilized as electrolytes to produce EW

Salts	Chemical formula	References
Ammonium molybdate	$(NH_4)_2MoO_4$	Hussien et al. (2018)
Copper sulfate	$CuSO_4$	Fallanaj et al. (2013)
EDTA-Ca	$C_{10}H_{12}CaN_2Na_2O_8$	Hussien et al. (2018)
EDTA-Fe	$C_{10}H_{12}FeN_2O_8$	
Magnesium chloride	$MgCl_2$	Buck et al. (2002)
Monopotassium phosphate	KH_2PO_4	Hussien et al. (2018)
Potassium bicarbonate	$KHCO_3$	
Potassium carbonate	K_2CO_3	Hussien et al. (2018) and Youssef and Hussien (2020)
Potassium chloride	KCl	Buck et al. (2002)
Potassium phosphate dibasic	K_2HPO_4	Fallanaj et al. (2013) and Hussien et al. (2018)
Potassium sorbate	$C_6H_7KO_2$	Fallanaj et al. (2013), Hussien et al. (2018) and Youssef and Hussien (2020)
Sodium bicarbonate	$NaHCO_3$	Fallanaj et al. (2013) and Hussien et al. (2018)
Sodium carbonate	Na_2CO_3	
Sodium chloride	NaCl	Park et al. (2001), Buck et al. (2002), Al-Haq et al. (2005), Hussien et al. (2018) and Youssef and Hussien (2020)
Sodium metabisulfite	$Na_2S_2O_5$	Hussien et al. (2018) and Youssef and Hussien (2020)
Sodium silicate	$(Na_2O)_x{\cdot}(SiO_2)$	Hussien et al. (2018)
Sodium sulfite	Na_2SO_3	Fallanaj et al. (2013)

to the anode donating electrons to generate oxygen (O_2), chlorine gas (Cl_2), hypochlorite ions (ClO^-), and hydrochloric acid (HCl), whereas positively charged ions, such as H^+ and Na^+, move to the cathode to accept electrons to generate hydrogen gas (H_2) and sodium hydroxide (NaOH) (Siddiqui 2018). When the electrolysis cell is separated by a septum, species produced on the anode stream result in an acidic solution of pH 2–3, an oxidation-reduction potential (ORP) higher than 1100 mV, and an active chlorine content (ACC) of 10–90 ppm. Species produced on the cathode stream result in a basic solution of pH 10–13 and an ORP of −800 to −900 mV. When the electrolysis cell is not separated by a septum, neutral electrolyzed water (NEW), with a ORP of 750–900 mV and pH of about 7 is produced, because hydroxide ions (OH^-) formed at the anode neutralizes the protons (H^+) produced at the cathode (Deza et al. 2007). Compared to other types of EW, NEW has a longer shelf-life under certain circumstances (Rahman et al. 2010a, b).

Indeed, EW is usually prepared on site just before use, but Len et al. (2002) demonstrated that AEW stored in a closed and dark environment remains stable. In particular, AEW rapidly decreased ORP releasing Cl_2 through the evolution of chlorine gas, thus rapidly reducing the biocidal effectiveness of the solutions. Len et al. (2002) observed a 100% loss of active chlorine and a 10% loss of ORP within a

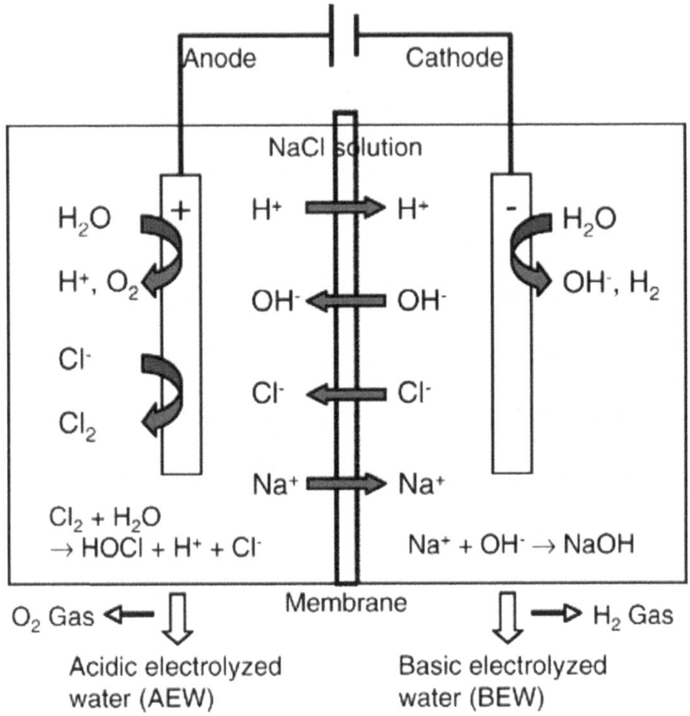

Fig. 12.1 Schematic representation of AEW and BEW generation using NaCl solution. (Hricova et al. 2008)

4-day period for AEW stored in an open dark container at 25 °C. In contrast, loss of chlorine oxidants and ORP of NEW was substantially lower, with only 5% decrease of active chlorine and no significant loss of ORP after 4 days in a closed dark container at 25 °C (Guentzel et al. 2010). Another way to preserve the effectiveness of EW is to convert it into ice cubes for later use (Koseki et al. 2002). Finally, slightly acidic electrolyzed water (SAEW) with a pH of 5.0–6.5 (Fig. 12.2) and an ORP of 800–900 mV is produced by electrolysis of diluted solution of HCl alone or in combination with NaCl in an EW generation equipment using an electrolysis chamber without the membrane (Forghani et al. 2015). SAEW usually has high ACC (up to 200 ppm) and for this reason can be used in a diluted form; its main free chlorine is HOCl (Fig. 12.3). The bactericidal activity of hypochlorous acid was 80 times greater than that of hypochlorite ion (ClO^-) for inactivating *Escherichia coli* at the same chlorine concentration and treatment time (Anonymous 1997). Therefore, SAEW may improve the bactericidal activity through maximizing the use of hypochlorous acid, thus reducing corrosion of surfaces and minimizing human health and safety issues side effects from off-gassing of Cl_2 (Guentzel et al. 2008).

EW is generally considered safer and less expensive than most traditional preservation methods. Various machines are manufactured around the world. The most

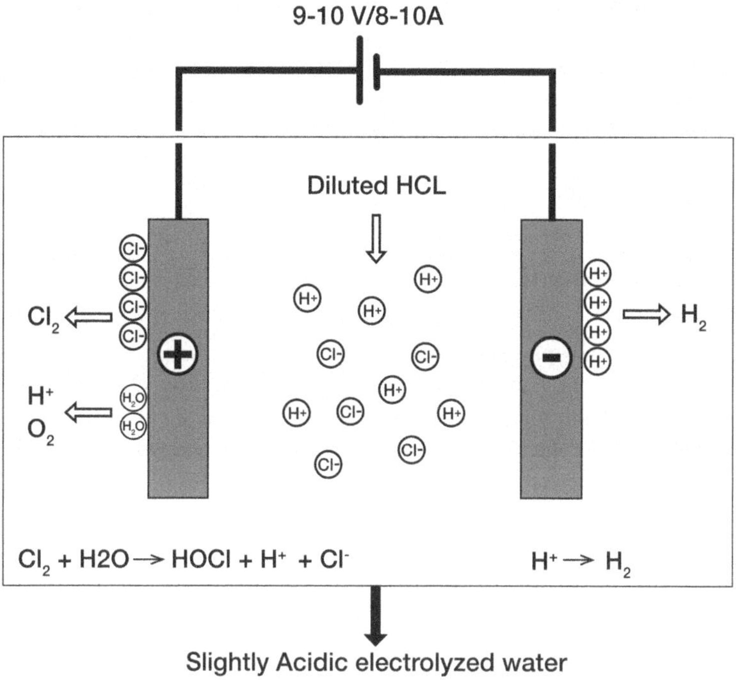

Fig. 12.2 Schematic representation of SAEW generation using diluted solution of HCl. (From Xuan and Ling 2019)

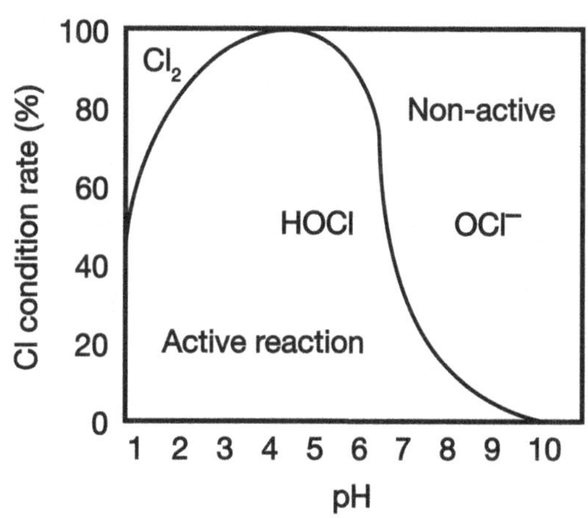

Fig. 12.3 Chlorine profile change with pH. The activities of Cl_2, HOCl, and OCl^- produced during the electrolysis process are pH dependent, being Cl_2 present at low pH, HOCl active at pH around 4.0–5.0, and OCl^- effective at high pH. (From Al-Haq et al. 2005)

common equipment used in published reports are ROX-20TA-U and ROX-10 WBE (Hoshizaki Electric Inc., Toyoake, Aichi, Japan), Remotex (Remote Co., Toshima-ku, Tokyo), model IKS 1005 (Mitsubishi Electric Engineering Co., Japan) (Al-Haq and Gómez-López 2012), and HRW-1500 (HuoRen-Jing-Chuang Medical Equipment

Co., Ltd., Beijing) (Chen et al. 2020). In Europe the most common equipment in published reports producing NEW are manufactured by Adamant Technology (SA, Switzerland) (López-Gálvez et al. 2012; Fallanaj et al. 2013, 2016, 2015), Denora Next (Milan, Italy), Best Life (China), ATS unique technologies BV (The Netherlands), ATS (Holambra, SP, Brasil) and more recently by Aqanat Limited (Coxwold, York, UK).

Factors Influencing the Effectiveness of EW

The electrode materials play an important role in the production of oxidant species in relation to the current, temperature, salt, and type of electrolysis (Martínez-Huitle and Brillas 2008). Traditionally, platinum is used as the anode in the EW generator. A descending order of electrode materials in terms of efficiency in producing active free chlorine was proposed by Rahman et al. (2016): Ti/IrO_2 > Ti/RuO_2 > $Ti/Pt–IrO_2$ > BDD (Boron-Doped Diamond) > Pt.

The influence of water hardness on the basic properties of EW has been reported by a few researchers (Pangloli and Hung 2013; Forghani et al. 2015). The authors reported that water hardness from 0 to 50 mg/L $CaCO_3$ increases the ACC and ORP levels of EW, while decreasing the pH; however, water hardness higher than 50 mg/L was observed to inhibit the inactivation of *E. coli* O157:H7 by EW. The mechanisms of how water hardness changes the bactericidal efficacy of EW still remains unclear and requires more investigations. Moreover, Forghani et al. (2015) highlighted that pre-heating water before EW generation allowed to increase the ACC and the biocide activity.

In a fresh produce processing plant, sanitizers generally are used in the presence of organic matter, such as produce debris, soils, and microorganisms present on fruit and vegetable surfaces, all of which reduce sanitizer efficacy. Oomori et al. (2000) reported that organic matter, including amino acids and proteins, potentially react with ACC and change it into the combined form. For instance, Li et al. (1996) observed that the reduction rate of *Bacillus subtilis* var. *niger* by EW exposure for 20 min decreased from 100 to 19.5% after adding 10% bovine serum albumin (BSA) to AEW. Indeed, organic matters might wrap target microorganisms and protect the outer structures of microbial cells from the attack of EW (Park et al. 2009; Virto et al. 2005).

In addition, the bactericidal effect of EW is thought to be better on smooth surfaces than on rough ones (Koseki et al. 2004; Park et al. 2009). For instance, Park et al. (2009) observed that the reduction of *E. coli* O157:H7, *Salmonella typhimurium*, and *Listeria monocytogenes* on the surface of a tomato by AEW exposure was higher than that on the surface of green onions. The authors explained that the smoother surface of tomatoes accelerated the activity of chlorine species in EW to better contact with microorganisms.

Effect on Plant Pathogens

The modes of action of electrochemical treatment of water are still not completely understood. Evidences suggest that a direct oxidation at the anode surface and indirect oxidation in the bulk solution by oxidants produced from the substances present in the water are responsible for the inactivation of microorganisms (Anglada et al. 2009). In *Aspergillus flavus* morphological changes occurred both in the conidia and mycelia, such as cell wall shrinkage, partial cracking, chipping, and holes (Fig. 12.4, Xiong et al. 2014). Chlorine compounds, pH, ORP, and their combination are considered the main factors involved in the antimicrobial activity (Al-Haq et al. 2005) and these are reported as the mode of action of many gaseous and aqueous oxidizing agents (Finnegan et al. 2010). However, since EW is active in a wide range of ORP and pH values and in some cases free chlorine is not generated, it is conceivable that its activity is related but not limited to these three factors. EW seems to induce higher sensitivity to active chlorine by sensitizing the outer membrane to the entry of HOCl (Park et al. 2004a). HOCl is considered the most active of the chlorine compounds (Mahmoud 2007) produced during electrolysis, penetrating cell membranes and producing hydroxyl radicals, which exert the antimicrobial activity through the oxidation of key metabolic compounds (Albrich et al. 1986; Barrette et al. 1989; Hurst et al. 1991; Hricova et al. 2008). HOCl can change bacterial respiration destroying the electron transport chains and affecting adenine nucleotide pool (Albrich et al. 1981). Chlorine is considered responsible of: (a) disruption of protein synthesis; (b) oxidative decarboxylation of amino acids to nitrites and aldehydes; (c) reactions with nucleic acids, purines, and pyrimidines; (d) unbalanced metabolism after the destruction of key enzymes; (e) induction of DNA lesions with the accompanying loss of DNA-transforming ability; (f) inhibition of oxygen uptake and oxidative phosphorylation, coupled with leakage of some macromolecules; (g) formation of toxic N-chlorine derivatives of cytosine; and (h) creation of chromosomal aberrations (Feliziani et al. 2016).

ORP is also involved in the mode of action of EW but its effect on the deactivation of microbes is controversial (McPherson 1993; Venkitanarayanan et al. 1999; Kim et al. 2000, 2001; Al-Haq et al. 2002; Liao et al. 2007). Aerobic bacteria grow mostly at ORP range + 200 to +800 mV, while anaerobic bacteria grow well at −700 to +200 mV (Jay 1996). The high ORP in the EW could cause the modification of metabolic fluxes and ATP production (Fig. 12.5), probably due to the change in the electron flow in cells (Huang et al. 2008).

Some authors suggested that the bacterial inactivation is primarily related to ORP and not to residual chlorine (Kim et al. 2000; Al-Haq et al. 2005). The high ORP of the solution affected fungi disrupting the outer membrane and facilitating the transfer of HOCl across the cell membrane, interfering on respiratory pathways (Liao et al. 2007). For example, it could cause damage to *E. coli* O157:H7, and attacked inner and outer membranes, causing necrosis of cells (Liao et al. 2007), with damage verified by microscopy (Feliciano et al. 2012). In case of NEW, produced by using diamond electrode and $NaHCO_3$ as electrolyte, the activity of free

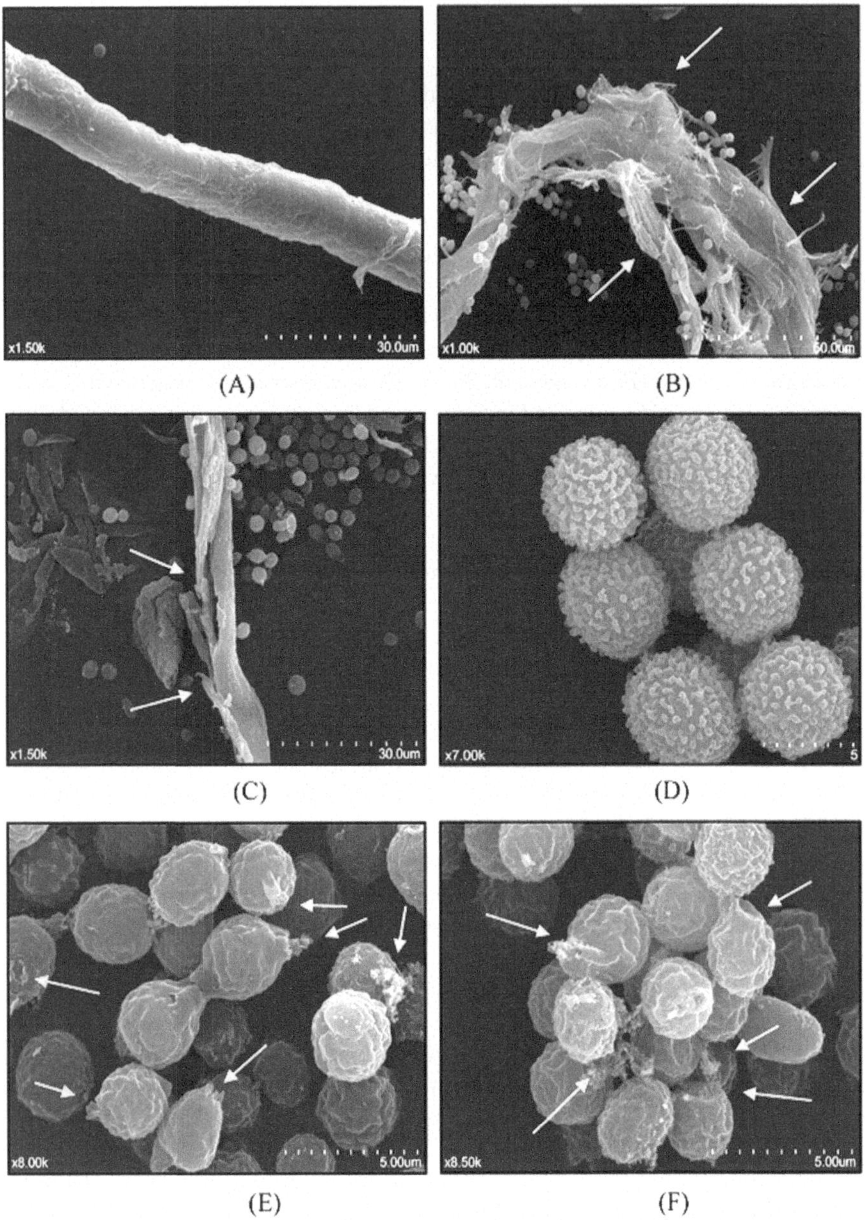

Fig. 12.4 Scanning electron photomicrographs of *Aspergillus flavus* conidia and mycelia. (**a**), normal mycelium; (**b**), mycelia treated with AEW; (**c**), mycelium treated with NEW; (**d**), normal conidia; (**e**), conidia treated with AEW; (**f**), conidia treated with NEW. Arrows show morphological changes in the conidia and mycelia, including cell wall shrinkage, partial cracking, chipping, and holes. Scale bars = (**a** and **c**), 30 μm; (**b**), 50 μm; (**d**–**f**), 5 μm. (From Xiong et al. 2014)

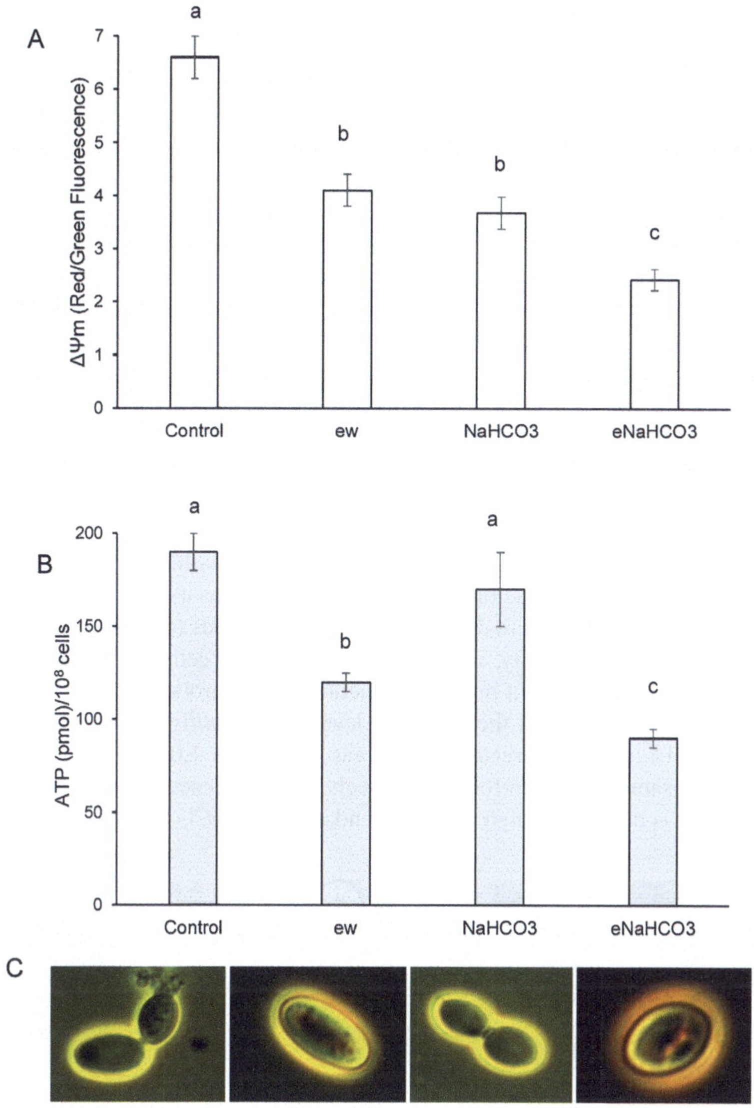

Fig. 12.5 Effect of electrolyzed water (ew), sodium bicarbonate (NaHCO3) and electrolyzed NaHCO3 (eNaHCO3) on mitochondrial membrane potential of Penicillium digitatum spores represented as red/green florescence ratio (**a**), and on ATP content (**b**). For each treatment representative images of stained spores under fluorescence microscopy are showed (**c**). Water was used as a control. (From Fallanaj et al. 2016)

chlorine was negligible and the pH little above the neutral value (Fallanaj et al. 2013); as reported also by Jeong et al. (2009), Fallanaj et al. (2013, 2016) ascribed the observed inactivation of *Penicillium* spp. population to electrochemical production of non-chlorine-based oxidants, such as hydrogen peroxide (H_2O_2), peroxymonocarbonate (HCO_4^-), and reactive oxygen species (ROS). In addition, the thin film-coated diamond electrode is known to produce by itself active oxygen species in a higher amount as compared to other anodes and, in presence of carbonate/bicarbonate-containing solutions, it can produce peroxycarbonate and derivatives, acting as strong disinfectants (Furuta et al. 2004). In addition, a paper by Fallanaj et al. (2016) demonstrated that electrolyzed $NaHCO_3$ solution, when applied in wounds nearby the ones inoculated with the pathogen, was able to control *P. digitatum* infections in citrus fruit and a significant up-regulation of defense-related genes coding the enzymes chitinase, peroxidase, and phenylalanine ammonia-lyase was observed in treated tissues. Differences in mycelia micromorphology of *Penicillium* species treated with various EW and untreated were displayed in Fig. 12.6 and Fig. 12.7 employing scanned electron microscopy (Youssef and Hussien 2020). Based on above results on the mode of action of EW, its antimicrobial effect derives from the combined action of pH, ORP, free chlorine, and other still unknown active substances (Huang et al. 2008); in addition, the induction of tissue resistance should be considered as another important aspect of the multiple mechanism of action of this technology (Fallanaj et al. 2016).

The pH also has its role in limiting the microbial growth; therefore, scientists also include it as one of the factors. Each microorganism has its own optimal growth range of pH; a low pH tends to destroy cell wall compounds (e.g. polysaccharides) and increase the permeability, resulting in the death of cell (McPherson 1993). Nevertheless, a low pH might not be sufficient to kill microbes, especially spores. Li et al. (1996) reported that the reduction level of *B. subtilis* var. *niger* can reach 100% after a 10-min AEW treatment, whereas it was only 1.06% for an HCl solution with the same pH. Therefore, most likely, the differences in effectiveness at different pHs is due to the high or low abundance of HOCl. In particular, at high

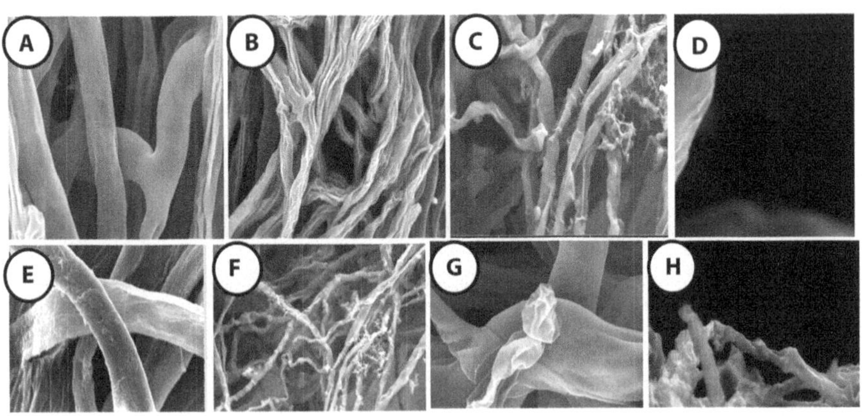

Fig. 12.6 Scanning electron microscope images of *Penicillium digitatum*-mycelium with free and linearly shaped hyphae (controls **a** and **e**). *P. digitatum*-mycelium in the presence of BEW generated by sodium metabisulphite (**b**), potassium sorbate (**c**) or potassium carbonate (**d**). *P. digitatum*-mycelium in the presence of AEW generated by sodium metabisulphite (**f**), potassium sorbate (**g**) or potassium carbonate (**h**). (From Youssef and Hussien 2020)

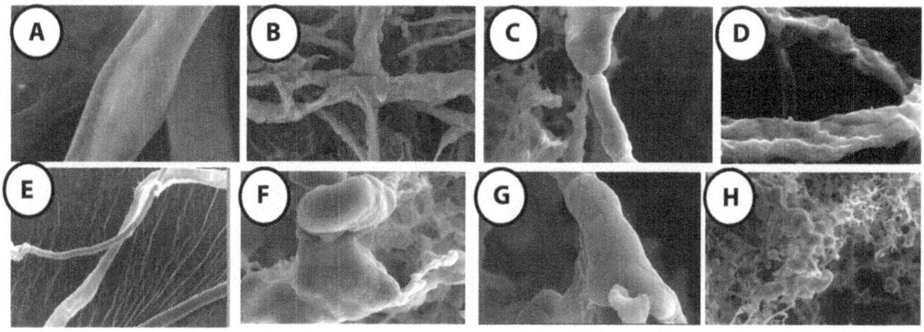

Fig. 12.7 Scanning electron microscope images of *Penicillium italicum*-mycelium with free and linearly shaped hyphae (controls **a** and **e**). *P. italicum*-mycelium in the presence of BEW generated by sodium metabisulphite (**b**), potassium sorbate (**c**) or potassium carbonate (**d**). *P. italicum*-mycelium in the presence of AEW generated by sodium metabisulphite (**f**), potassium sorbate (**g**) or potassium carbonate (**h**). (From Youssef and Hussien 2020)

pHs, the concentration of HOCl decreased, reflecting its dissociation to H^+ and OCl^- (Johnson and Melbourne 1996; White 1998).

Based on above literature there is no a consensus about EW mode of action against microorganisms (Table 12.2), but a lot of theories exist. Likely, multiple mechanisms are responsible of EW biocidal activity and this is theoretically confirmed by the absence of pathogen resistance.

Effect on Microbial Toxins

A study conducted by Audenaert et al. (2012) demonstrated that EW has potential to control *Fusarium* spp. in wheat grains during transport and storage although sublethal concentrations can result in increased deoxynivalenol (DON) biosynthesis. According to Zhang et al. (2012), soaking contaminated peanuts in an EW solution, the content of aflatoxin B1 (AFB_1) decreased of about 85%. Moreover, they reported better results with AEW, suggesting a stronger decontamination effect of HClO than ClO^-. On the same line, Suzuki et al. (2002) reported a strong reduction of the mutagenesis effect of AFB_1 against *Salmonella typhimurium* TA-98 and TA-100 strains after the exposure of the toxin to the AEW.

Effect on Plants

Considering a holistic approach to the crop protection, it should be taken into account the effect of EW not only against the pathogens, but also on the crop. It has been demonstrated that SAEW inhibited the growth of broccoli sprouts, but increased sulforaphane content (Li et al. 2018). Another study conducted on Chinese

Table 12.2 Studies conducted on the effect of EW water on various microorganisms

Microbial species	References
Acidovorax avenae subsp. *citrulli*	Buck et al. (2002)
Alternaria sp.	
Alternaria panax	
Aspergillus flavus	Buck et al. (2002) and Xiong et al. (2014)
Aspergillus spp.	Suzuki et al. (2002)
Botryosphaeria berengeriana	Al-Haq et al. (2002)
Botrytis allii	Buck et al. (2002)
Botrytis cinerea	Buck et al. (2002), Guentzel et al. (2010), Guentzel et al. (2011) and Youssef et al. (2018)
Cladosporium sp.	Buck et al. (2002)
Colletotrichum sp.	
Colletotrichum fructicola	Hirayama et al. (2016)
Curvularia lunata	Buck et al. (2002)
Didymella bryonaie	
Epicoccum nigrum	
Erwinia chrysanthemi	
Fusarium sp.	
Fusarium moniliforme	
Helminthosporium sp.	
Monilinia fructicola	Al-Haq et al. (2001), Buck et al. (2002) and Guentzel et al. (2010)
Pantoea ananatis	Buck et al. (2002)
Penicillium digitatum	Whangchai et al. (2010) and Fallanaj et al. (2013, 2016)
Penicillium expansum	Okull and Laborde (2004)
Penicillium italicum	Whangchai et al. (2010) and Fallanaj et al. (2013, 2016)
Penicillium ulaiense	Hussien et al. (2018)
Pestalotia sp.	Buck et al. (2002)
Phomopsis longicolla	
Pseudomonas syringae pv. *syringae*	
Pseudomonas syringae pv. *glycinea*	
Pseudomonas spp.	Fallanaj et al. (2015)
Pseudomonas fluorescens	Pinto et al. (2015)
Pseudomonas marginalis	
Pseudomonas syringae	
Rhodosporidium toruloides	Buck et al. (2002)
Sphaerotheca fuliginea	Fujiwara et al. (2009)

(continued)

Table 12.2 (continued)

Microbial species	References
Stagonospora nodorum	Buck et al. (2002)
Thielaviopsis basicola	
Tilletia indica	Bonde et al. (1999)
Trichoderma spirale	Buck et al. (2002)
Total bacteria	Ding et al. (2015)
Psychrophilic bacteria	Gómez-López et al. (2013)
Total aerobic bacteria	Koide et al. (2009), Rahman et al. (2010a, b), Hao et al. (2011a, b, 2015a, b), Zhang et al. (2016a, b), Li et al. (2017) and Tango et al. (2017)
Various fungi and bacteria	Koseki et al. (2004)
Yeasts and molds	Koide et al. (2009), Rahman et al. (2010a, b), Hao et al. (2011a, b, 2015a, b), Navarro-Rico et al. (2014), Ding et al. (2015), Zhang et al. (2016a, b) and Li et al. (2017)

cabbage highlighted that foliar application of EW solution could enhance the photosynthetic rate, leaf number, and yield; instead, root applications could increase the content of vitamin C (Hou et al. 2011). The application of EW on harvested sugarcane during summer months showed relatively less decline in Commercial Cane Sugar (CCS), sucrose, and purity of juice compared to untreated and water-treated control (Solomon and Singh 2009).

The effect of EW water on respiration rate is variable; it can increase, decrease, or remain unchanged. After AEW treatment, it has been reported that the respiration rate increased in lettuce and cabbage (Koseki and Itoh 2002). In contrast, an EW treatment on leek, white cabbage, and mizuna baby leaf, reduced the respiration rate significantly (Vandekinderen et al. 2009a, b). Finally, a NEW treatment on grated carrots and iceberg lettuce did not influence the respiration rate (Vandekinderen et al. 2008; Vandekinderen et al. 2009c). Moreover, depending on the fruits or vegetables and especially for "minimally processed" produce, the treatment with EW could have some effects on the nutritional and phytochemical composition, due to the oxidation nature of the EW and/or by leaching of substances from vegetable tissue due to water-vegetable surface contact (Al-Haq and Gómez-López 2012). Other researchers showed that changes in respiration rate during cold storage of cabbages and broccoli could be avoided by EW (Gómez-López et al. 2007; Navarro-Rico et al. 2014).

Electrolyzed Water and Quality of Produce

It is well known that fruit and vegetable quality is becoming more relevant than market price to most of the consumers. Unfortunately, most of research accounts have tested the effect of treatment on pathogens, while any possible negative consequence on fruit quality is not often acknowledged. Few studies were performed to investigate the effect of EW on produce quality. Youssef and Hussien (2020) summarized that neither BEW nor AEW have any harmful effect in terms of citrus quality including weight loss, total soluble solids, citric acid, ascorbic acid, pH and color index. Some scientists found no statistical difference in color index of lettuce, broccoli, strawberry, and date palm before and after EW treatment (Park et al. 2001; Hung et al. 2010a, b; Jemni et al. 2014). Also, EW application proved to have no harmful effect on iceberg lettuce and white cabbage quality with regard to vitamin C loss (Vandekinderen et al. 2009a, b). In addition, the use of EW had no obvious effect on both titratable acidity and pH in the case of date fruit and strawberry (Hung et al. 2010a, b; Bessi et al. 2014).

Effect on Removing Pesticide Residues

To ensure a sustainable production of fruits, vegetables, and grains, the farmers use a variety of pesticides to protect the crops from insects, mites, fungi, bacteria, weeds, etc. However, when humans and animals consume foods with pesticide residues, they can cause cumulative poisoning effects. Several physical, chemical, and biological methods including adsorption, oxidation, catalytic degradation, membrane filtration, and biological treatment have been developed in order to remove/inactivate pesticide residues. The use of EW as potential tool to remove pesticide residues, has been evaluated during the last years (Hao et al. 2011b; Sung et al. 2011, 2012; Wuyun 2011; Hao and Li 2006; Luo et al. 2014; Liu et al. 2015; Hu et al. 2016; Han et al. 2017; Qi et al. 2018), as shown in Table 12.3.

They showed that a higher ACC and a longer treatment time leaded to greater reductions of pesticide residues. Moreover, the effectiveness was dependent on the chemical proprieties of the pesticides. For example, organophosphorus pesticides (e.g. dimethoate, chlorpyrifos, etc.), containing P=S double bonds and P–S or P–O single bonds, are easily attacked by chlorin (Deborde and von Gunten 2008) and hence can be degraded by the available chlorine in EW (Qi et al. 2018). In addition, a nucleophilic reaction has been reported to occur under acidic or alkaline conditions with the break of the double bond because of AEW low pH and high ORP value, whereas BEW with its high pH has proved to have a good emulsifying property (Wang and Han 2019). In summary, the EW has an obvious effect on the removal of pesticide residues on food without a significant decrease in quality (Wang and Han 2019).

Table 12.3 Studies conducted on the effect of EW water in removing pesticide residues from produce

Pesticides	Samples	Percentage reduction[a] (%)	References
Chlorpyrifos	Chinese cabbage	50–70	Sung et al. (2012)
Prothiofos	Chinese cabbage	50–75	
Deltamethrin	Chinese cabbage	40–80	
Chlorpyrifos	*yuja*	70	Sung et al. (2011)
Prothiofos	*yuja*	70	
Spirodiclofen	*yuja*	72	
Deltamethrin	*yuja*	82	
Benomyl	*yuja*	97	
Thiophanate-methyl	*yuja*	98	
Acequinocyl	*yuja*	82	
Isoprocarb	Cowpea	17–85	Han et al. (2017)
Chlorpyrifos	Cowpea	10–60	
Bifenthrin	Cowpea	9–48	
Beta-cypermethrin	Cowpea	6–56	
Difenoconazole	Cowpea	9–68	
Azoxystrobin	Cowpea	8–75	Han et al. (2017)
Dimethoate	Apple	35–81	Wuyun (2011)
Chlorpyrifos	Apple	27–72	
Parathion	Apple	31–77	
Dimethoate	Rape	>60 to >70	
Dimethoate	Tomato and beans	>50	
Chlorpyrifos	Rape	>50	
Chlorpyrifos	Tomato and beans	>50	
Parathion	Rape and tomato	>60	
Parathion	Beans	>50	
Acephate	Rape	82 to >90	Hao and Li (2006)
Lambda-cyhalothrin	Apple	40–90	Liu et al. (2015)
Dimethoate	Leek	40–79	Hu et al. (2016)
Chlorpyrifos	Leek	40–73	
Phoxim	Cabbage	92	Luo et al. (2014)

[a]Percentage reduction range depend on pH, ACC, ORP, and treatment time

Advantages of EW

The on-site production of the EW, whatever the use, represents a great advantage because there are no chemicals to purchase or store, except for an inexpensive salt (NaCl or others), eliminating the need for purchasing, transporting, storing, preparing and using traditional chemicals. EW has minimal impact on the environment (Koseki et al. 2002); particularly, NEW and BEW are safe for the environment and the operators since little chlorine is released to the air. If non-chlorine salts (e.g. $NaHCO_3$) are used as electrolytes, health concerns with regard to chlorine in the air and in water are avoided and, consequently, the formation of chlorinated organic compounds including chloramines (NH_2Cl), dichloramines ($NHCl_2$), and trichloromethanes ($HCCl_3$). These are respiratory irritants suspected to be carcinogenic (Roberts and Reymond 1994; Fallanaj et al. 2013; Citizens Concerned About Chloramine 2019. EW reverts to normal water after use, and its effectiveness has been verified within a large pH range (Park et al. 2004b). Since EW has multiple mechanisms of action, it is quite unlikely that resistance in target microorganisms will develop (Al-Haq et al. 2005).

After the initial cost of the apparatus for electrolysis, operational expenses become minimal (Bonde et al. 1999) and the capital cost of the on-site apparatus can often be recovered in less than a few years.

Indeed, in USA the unit cost per kilogram of electronically generated chlorine is significantly cheaper than liquefied chlorine gas, sodium hypochlorite solution, dry calcium hypochlorite, and cyanurate-based (TCIA) tablets (Grech and Rijkenberg 1992). The raw materials, water and sodium chloride, are found virtually everywhere (Venczel et al. 1997). Its use reduces the hazards associated with handling, transportation, and storage of concentrated chlorine solution (Nakagawara et al. 1998). The biocidal capacity of EW as compared to traditional chemical solutions permits the use of low dose rates, reducing the risk for environmental impact and the solutions should be less corrosive than alternate products. Lastly, the use of EW on various food commodities did not negatively affect the organoleptic properties, color, scent, flavor, or texture (Al-Haq et al. 2005; Hricova et al. 2008; Huang et al. 2008).

Disadvantages of EW

Strongly acidic EW and free chlorine content may be corrosive to some metals and may induce synthetic resin degradation (Tanaka et al. 1999), and hazardous chlorinated by-products can be produced. Its effectiveness may be hindered by the presence of organic substances (Oomori et al. 2000); its antimicrobial potential could be loosed quickly, once the apparatus is switched-off (Kiura et al. 2002). Depending on the electrolyte used and the pH (e.g. in AEW), pungent chlorine gas is formed that can cause discomfort to operators (Al-Haq et al. 2005). Excessive chlorine can be potentially toxic for plant produce (Grech and Rijkenberg 1992). AEW can induce phytotoxicity; for example, white spots and slight necrosis were observed on flowers and leaf edges of some ornamental bedding plants following an AEW foliar spray (Buck et al. 2003). A drawback could be also the need to switch-on the apparatus one or few hours before utilization to allow the bulk of water to become rich in antimicrobial oxidizing species (Fallanaj et al. 2013).

Conclusion

Disinfection of fresh produce and storage facilities is generally an important requirement for postharvest decay management. The applicability of disinfectants to control postharvest fruit decay depends on many aspects, i.e. on the fresh produce, the orientation towards organic or conventional agriculture, the time of the produce storage, the characteristics of the postharvest facilities, the possibilities to integrate the disinfection operations with other technologies and the know-how of the staff. In general, EW is characterized by a low impact on the environment and the operators, leaving no toxic residues or eliminating them on the food matrix. Because EW have multiple mechanisms of action it is quite unlikely that resistance in target microorganisms could develop. In view of the potential benefits to extend the storage period of fruit provided by the disinfectant agents, further studies could optimize their integration into current practices of postharvest manipulation.

References

Albrich JM, McCarthy CA, Hurs JK (1981) Biological reactivity of hypochlorous acid: implications for microbicidal mechanisms of leukocyte myeloperoxidase. Proc Natl Acad Sci U S A 78(1):210–214

Albrich JM, Gilbaugh JH, Callahan KB et al (1986) Effects of the putritive neutrophil– generated toxin, hypochlorous acid, on membrane permeability and transport systems of *Escherichia coli*. J Clin Invest 78:177–184

Al-Haq MI, Gómez-López VM (2012) Electrolyzed oxidizing water. In: Gómez-López VM (ed) Decontamination of fresh and minimally processed produce, 1st edn. Wiley-Blackwell, Hoboken, pp 135–164

Al-Haq MI, Seo Y, Oshita S et al (2001) Fungicidal effectiveness of electrolyzed oxidizing water on post harvest brown rot of peach. Hortic Sci 36:1310–1314

Al-Haq MI, Seo Y, Oshita S et al (2002) Disinfection effects of electrolyzed oxidizing water on suppressing fruit rot of pear caused by *Botryosphaeria berengeriana*. Food Res Int 35:657–664

Al-Haq MI, Sugiyama J, Isobe S (2005) Applications of electrolyzed water in agriculture and food industries. Food Sci Technol Res 11:135–150

Anglada A, Urtiaga A, Ortiz I (2009) Contributions of electrochemical oxidation to waste-water treatment: fundamentals and review of applications. J Chem Technol Biotechnol 84:1747–1755

Anonymous (1997) Principle of formation of electrolytic water. Hoshizaki Electic Co., Ltd., Sakae, Toyoake

Audenaert K, Monbaliu S, Deschuyffeleer N et al (2012) Neutralized electrolyzed water efficiently reduces *Fusarium* spp. *in vitro* and on wheat kernels but can trigger deoxynivalenol (DON) biosynthesis. Food Control 23(2):515–521

Barrette WCJ, Hannum DM, Wheeler WD et al (1989) General mechanism for the bacterial toxicity of hypochlorous acid: abolition of ATP production. Biochemistry-US 28:9172–9178

Bessi H, Debbabi H, Grissa K et al (2014) Microbial reduction and quality of stored date fruits treated by electrolyzed water. J Food Qual 37:42–49

Bonde MR, Nester SE, Khayat A et al (1999) Comparison of effects of acidic electrolyzed water and NaOCl on *Tilletia indica* teliospore germination. Plant Dis 83:627–632

Buck JW, Van Iersel MW, Oetting RD et al (2002) In vitro fungicidal activity of acidic electrolyzed oxidizing water. Plant Dis 86(3):278–281

Buck JW, Van Iersel MW, Oetting RD et al (2003) Evaluation of acidic electrolyzed water for phytotoxic symptoms on foliage and flowers of bedding plants. Crop Prot 22:73–77

Chen Y, Xie H, Tang J et al (2020) Effects of acidic electrolyzed water treatment on storability, quality attributes and nutritive properties of longan fruit during storage. Food Chem 320:126641

Citizens Concerned About Chloramine (2019) http://www.chloramine.org/. Accessed 15 Jan 2020

Deborde M, Von Gunten URS (2008) Reactions of chlorine with inorganic and organic compounds during water treatment kinetics and mechanisms: a critical review. Water Res 42:13–51

Deza M, Araujo M, Garrido M (2007) Efficacy of neutral electrolyzed water to inactivate *Escherichia coli, Listeria monocytogenes, Pseudomonas aeruginosa,* and *Staphylococcus aureus* on plastic and wooden kitchen cutting boards. J Food Prot 70(1):102–108

Ding T, Ge Z, Shi J et al (2015) Impact of slightly acidic electrolyzed water (SAEW) and ultrasound on microbial loads and quality of fresh fruits. Food Sci Technol 60(2):1195–1199

Fallanaj F, Sanzani SM, Zavanella C et al (2013) Salt addition improves the control of citrus postharvest diseases using electrolysis with conductive diamond electrodes. J Plant Pathol 95(2):373–383

Fallanaj F, Sanzani SM, Youssef K et al (2015) A new perspective in controlling postharvest citrus rots: the use of electrolyzed water. Acta Hortic 1065:1599–1605

Fallanaj F, Ippolito A, Ligorio A et al (2016) Electrolyzed sodium bicarbonate inhibits *Penicillium digitatum* and induces defence responses against green mould in citrus fruit. Postharvest Biol Technol 115:18–29

Feliciano L, Lee J, Pascall MA (2012) Transmission electron microscopic analysis showing structural changes to bacterial cells treated with electrolyzed water and an acidic sanitizer. J Food Sci 77(4):M182–M187

Feliziani E, Lichter A, Smilanick JL et al (2016) Disinfecting agents for controlling fruit and vegetable diseases after harvest. Postharvest Biol Techol 122:53–69

Finnegan M, Linley E, Denyer SP et al (2010) Mode of action of hydrogen peroxide and other oxidizing agents: differences between liquid and gas forms. J Antimicrob Chemoth 65:2108–2115

Forghani F, Park JH, Oh DH (2015) Effect of water hardness on the production and microbicidal efficacy of slightly acidic electrolyzed water. Food Microbiol 48:28–34

Fujiwara K, Fujii T, Park JS (2009) Comparison of foliar spray efficacy of electrolytically ozonated water and acidic electrolyzed oxidizing water for controlling powdery mildew infection on cucumber leaves. Ozone Sci Eng 31:10–14

Furuta T, Tanaka H, Nishiki Y et al (2004) *Legionella* inactivation with diamond electrodes. Diam Relat Mater 13:2016–2019

Gómez-López VM, Ragaert P, Ryckeboer J et al (2007) Shelf-life of minimally processed cabbage treated with neutral electrolysed oxidising water and stored under equilibrium modified atmosphere. Int J Food Microbiol 117(1):91–98

Gómez-López VM, Marín A, Medina-Martínez MS et al (2013) Generation of trihalomethanes with chlorine-based sanitizers and impact on microbial, nutritional and sensory quality of baby spinach. Postharvest Biol Technol 85:210–217

Grech NM, Rijkenberg FHJ (1992) Injection of electronically generated chlorine into citrus micro-irrigation systems for the control of certain waterborne root pathogens. Plant Dis 76:446–457

Guentzel JL, Lam KL, Callan MA et al (2008) Reduction of bacteria on spinach, lettuce, and surfaces in food service areas using neutral electrolyzed oxidizing water. Food Microbiol 25:36–41

Guentzel JL, Lam KL, Callan MA et al (2010) Postharvest management of gray mould and brown rot on surfaces of peaches and grapes using electrolyzed oxidizing water. Int J Food Microbiol 143:54–60

Guentzel JL, Callan MA, Emmons SA et al (2011) Evaluation of electrolyzed oxidizing water for phytotoxic effects and pre-harvest management of gray mold disease on strawberry plants. Crop Prot 30:1274–1279

Han YT, Song L, An QS et al (2017) Removal of six pesticide residues in cowpea with alkaline electrolysed water. J Sci Food Agric 97:2333–2338
Hao JX, Li LT (2006) Study on the removal of pesticide residue of vegetables by electrolyzed functional water. Sci Technol Food Ind 5:164–166
Hao J, Liu H, Liu R et al (2011a) Efficacy of slightly acidic electrolyzed water (SAEW) for reducing microbial contamination on fresh-cut Cilantro. J Food Saf 31(1):28–34
Hao J, Liu H, Chen T et al (2011b) Reduction of pesticide residues on fresh vegetables with electrolyzed water treatment. J Food Sci 76(4):C520–C524
Hao J, Li H, Wan Y et al (2015a) Combined effect of acidic electrolyzed water (AcEW) and alkaline electrolyzed water (AlEW) on the microbial reduction of fresh-cut cilantro. Food Control 50:699–704
Hao J, Li H, Wan Y et al (2015b) Effect of slightly acidic electrolyzed water (SAEW) treatment on the microbial reduction and storage quality of fresh-cut Cilantro. J Food Process Pres 39(6):559–566
Hirayama Y, Asano S, Watanabe K et al (2016) Control of *Colletotrichum fructicola* on strawberry with a foliar spray of neutral electrolyzed water through an overhead irrigation system. J Gen Plant Pathol 82:186–189
Hou M, Gao J, Deng L et al (2011) Effect of electrolyzed water on growth and development, nutritional quality of Chinese cabbage. Hubei Agric Sci 7(50):1342–1346
Hricova D, Stephan R, Zweifel C (2008) Electrolyzed water and its application in the food industry. J Food Prot 71(9):1934–1947
Hu ZH, Wu TJ, Wan YF et al (2016) Study on the removal of dimethoate and chlorpyrifos in leek by slightly acidic electrolyzed water. Sci Technol Food Ind 37(01):49–52
Huang YR, Hung YC, Hsu SY et al (2008) Application of electrolyzed water in the food industry. Food Control 19:329–345
Hung YC, Bailly D, Kim C et al (2010a) Effect of electrolyzed oxidizing water and chlorinated water treatments on strawberry and broccoli quality. J Food Qual 33(5):578–598
Hung YC, Tilly P, Kim C (2010b) Efficacy of electrolyzed oxidizing (Eo) water and chlorinated water for inactivation of *Escherichia coli* O157:H7 on strawberries and broccoli. J Food Qual 33(5):559–577
Hurst JK, Barrette WCJ, Michel BR et al (1991) Hypochlorous acid and myeleperoxidasecatalyzed oxidation of iron-sulfur clusters in bacterial respiratory dehydrogenase. Eur J Biochem 202:1275–1282
Hussien A, Al-Essawy A, Rehab A et al (2017) Preliminary investigation of alkaline and acidic electrolysed water to control *Penicillium* species of citrus. Citrus Res Technol 38(2):175–183
Hussien A, Ahmed Y, Al-Essawy AH et al (2018) Evaluation of different salt-amended electrolysed water to control postharvest moulds of citrus. Trop Plant Pathol 43(1):10–20
Jay JM (1996) Modern food microbiology, 5th edn. Aspen Pub, Frederick, pp 48–49
Jemni M, Gomez P, Souza M et al (2014) Combined effect of UV-C, ozone and electrolyzed water for keeping overall quality of date palm. Food Sci Technol 59(2):649–655
Jeong J, Kim C, Yoon J (2009) The effect of electrode material on the generation of oxidants and microbial inactivation in the electrochemical disinfection processes. Water Res 43:895–901
Johnson M, Melbourne P (1996) Photolytic spectroscopic quantification of residual chlorine in potable waters. Analyst 121:1075–1078
Kim C, Hung Y, Brackett RE (2000) Roles of oxidation-reduction potential in electrolyzed oxidizing and chemically modified water for the inactivation of food-related pathogens. J Food Prot 63:19–24
Kim C, Hung Y-C, Brackett RE et al (2001) Inactivation of *Listeria monocytogenes* biofilms by electrolyzed oxidizing water. J Food Process Preserv 25:91–100
Kiura H, Sano K, Morimatsu S et al (2002) Bactericidal activity of electrolyzed acid water from solution containing sodium chloride at low concentration, in comparison with that at high concentration. J Microbiol Meth 49(3):285–293

Koide S, Takeda J, Shi J et al (2009) Disinfection efficacy of slightly acidic electrolyzed water on fresh cut cabbage. Food Control 20(3):294–297
Koseki S, Itoh K (2002) Effect of nitrogen gas packaging on the quality and microbial growth of freshcut vegetables under low temperature. J Food Prot 65(2):326–332
Koseki S, Fujiwara K, Itoh K (2002) Decontamination effect of frozen acidic electrolyzed water on lettuce. J Food Prot 65:411–414
Koseki S, Yoshida K, Isobe S et al (2004) Efficacy of acidic electrolyzed water for microbial decontamination of cucumbers and strawberries. J Food Prot 67(6):1247–1251
Kunina LA (1967) From experience in the electrolytic decontamination of drinking water. Gig Sanit 32:100–101
Lee JH, Rhee P, Kim JH et al (2004) Efficacy of electrolyzed acid water in reprocessing patient-used flexible upper endoscopes: comparison with 2% alkaline glutaraldehyde. J Gastroenterol Hepatol 19:897–903
Len SV, Hung YC, Chung D et al (2002) Effects of storage conditions and pH on chlorine loss in electrolyzed oxidizing (EO) water. J Agric Food Chem 50(1):209–212
Li XW, Sun SH, Li T (1996) Preliminary study of microbiocide effect and its mechanism of electrolyzed oxidizing water. Zhonghua liu xing bing xue za zhi = Zhonghua liuxingbingxue zazhi 17(2):95
Li J, Ding T, Liao X et al (2017) Synergetic effects of ultrasound and slightly acidic electrolyzed water against *Staphylococcus aureus* evaluated by flow cytometry and electron microscopy. Ultrason Sonochem 38:711–719
Li L, Hao J, Song S et al (2018) Effect of slightly acidic electrolyzed water on bioactive compounds and morphology of broccoli sprouts. Food Res Int 105:102–109
Liao LB, Chen WM, Xiao XM (2007) The generation and inactivation mechanism of oxidationreduction potential of electrolyzed oxidizing water. J Food Eng 78:1326–1332
Liu HJ, Li RZ, Su DH et al (2015) Degradation of Lambda-cyhalothrin in fruits and vegetable by alkaline electrolyzed water. Food Sci Technol 40(2):123–127
López-Gálvez F, Posada-Izquierdo GD, Selma MV et al (2012) Electrochemical disinfection: an efficient treatment to inactivate *Escherichia coli* O157: H7 in process wash water containing organic matter. Food Microbiol 30(1):146–156
Luo Q, Zu YH, Shi KQ et al (2014) Study on the reduction effect of pesticide residues of vegetables with slightly acidic electrolyzed water. J Food Saf Qual 5(11):3657–3663
Mahmoud BS (2007) Electrolyzed water: a new technology for food decontamination - a review. Dtsch Lebensmitt Rundsch 103(5):212–221
Martínez-Huitle CA, Brillas E (2008) Electrochemical alternatives for drinking water disinfection. Angew Chem Int Edit 47(11):1998–2005
McPherson LL (1993) Understanding ORP's in the disinfection process. Water Eng Manage 140:29–31
Nakagawara S, Goto T, Nara M et al (1998) Spectroscopic characterization and the pH dependence of bactericidal activity of the aqueous chlorine solution. Anal Sci 14:691–698
Nakajima N, Nakano T, Harada F et al (2004) Evaluation of disinfective potential of reactivated free chlorine in pooled tap water by electrolysis. J Microbiol Meth 57(2):163–173
Navarro-Rico J, Artes-Hernandez F, Gomez PA et al (2014) Neutral and acidic electrolysed water kept microbial quality and health promoting compounds of fresh-cut broccoli throughout shelf life. Innov Food Sci Emerg Technol 21:74–81
Okull DO, Laborde LF (2004) Activity of electrolyzed oxidizing water against *Penicillium expansum* on suspension and on wounded apples. J Food Sci 69:23–27
Oomori T, Oka T, Inuta T et al (2000) The efficiency of disinfection of acidic electrolyzed water in the presence of organic materials. Anal Sci 16:465–469
Pangloli P, Hung YC (2013) Effects of water hardness and pH on efficacy of chlorine-based sanitizers for inactivating *Escherichia coli* O157:H7 and *Listeria monocytogenes*. Food Control 32(2):626–631

Park CM, Hung YC, Doyle MP et al (2001) Pathogen reduction and quality of lettuce treated with electrolyzed oxidizing and acidified chlorinated water. J Food Sci 66:1368–1372

Park SG, Lim HJ, Kim JK et al (2004a) Microbiocidal effect of *E. coli* with ozone solution generated by BDD electrode. http://www.electrochem.org/dl/ma/206/pdfs/2185.pdf. Accessed 24 Jan 2020

Park H, Hung YC, Chung D (2004b) Effects of chlorine and pH on efficacy of electrolyzed water for inactivating *Escherichia coli* 157:H7 and *Listeria monocytogenes*. Int J Food Microbiol 91:13–18

Park EJ, Alexander E, Taylor GA et al (2009) The decontaminative effects of acidic electrolyzed water for *Escherichia coli* O157:H7, *Salmonella typhimurium*, and *Listeria monocytogenes* on green onions and tomatoes with differing organic demands. Food Microbiol 26:386–390

Pinto L, Ippolito A, Baruzzi F (2015) Control of spoiler *Pseudomonas* spp. on fresh cut vegetables by neutral electrolyzed water. Food Microbiol 50:102–108

Qi H, Huang Q, Hung YC (2018) Effectiveness of electrolyzed oxidizing water treatment in removing pesticide residues and its effect on produce quality. Food Chem 239:561–568

Rahman SME, Ding T, Oh DW (2010a) Inactivation effect of newly developed low concentration electrolyzed water and other sanitizers against microorganisms on spinach. Food Control 21(10):1383–1387

Rahman SME, Jin YG, Oh DW (2010b) Combined effects of alkaline electrolyzed water and citric acid with mild heat to control microorganisms on cabbage. J Food Sci 75:M111–M115

Rahman SME, Khan I, Oh DH (2016) Electrolyzed water as a novel sanitizer in the food industry: current trends and future perspectives. Compr Rev Food Sci Food Saf 15:471–490

Roberts RG, Reymond ST (1994) Chlorine dioxide for reduction of postharvest pathogen inoculum during handling of tree fruits. Appl Environ Microbiol 60:2864–2868

Sargent SA, Ritenour MA, Brecht JK (2000) Handling, cooling and sanitation techniques for maintaining postharvest quality. University of Florida Cooperative Extension Service, Institute of Food and Agriculture Sciences, EDIS

Siddiqui MW (2018) Postharvest disinfection of fruits and vegetables. Academic, Elsevier

Solomon S, Singh P (2009) Efficacy of electrolyzed water to minimize postharvest sucrose losses in sugarcane. Sugar Tech 11(2):228–230

Sung JM, Kwon KH, Kim JH et al (2011) Effect of washing treatments on pesticide residues and antioxidant compounds in *Yuja* (*Citrus junos* Sieb ex Tanaka). Food Sci Biotechnol 20(3):767–773

Sung JM, Park KJ, Lim JH et al (2012) Removal effects of microorganism and pesticide residues on Chinese cabbages by electrolyzed water washing. Korean J Food Sci Technol 44(5):628–633

Suzuki T, Noro T, Kawamura Y et al (2002) Decontamination of aflatoxin-forming fungus and elimination of aflatoxin mutagenicity with electrolyzed NaCl anode solution. J Agric Food Chem 50:633–641

Tanaka N, Fujisawa T, Daimon T et al (1999) The effect of electrolyzed strong acid aqueous solution on hemodialysis equipment. Artif Organs 23:1055–1062

Tango C, Khan I, Ngnitcho K et al (2017) Slightly acidic electrolyzed water combined with chemical and physical treatments to decontaminate bacteria on fresh fruits. Food Microbiol 67: 97–105

Vandekinderen I, Van Camp J, Devlieghere F et al (2008) Effect of decontamination agents on the microbial population, sensory quality, and nutrient content of grated carrots (*Daucus carota* L.). J Agric Food Chem 56:5723–5731

Vandekinderen I, Van Camp J, Devlieghere F et al (2009a) Evaluation of the use of decontamination agents during fresh-cut leek processing and quantification of their effect on its quality by means of multidisciplinary approach. Innov Food Sci Emerg Technol 10:363–373

Vandekinderen I, Van Camp J, Devlieghere F et al (2009b) Effect of decontamination on the microbial load, the sensory quality and the nutrient retention of ready-to-eat white cabbage. Eur Food Res Technol 229:443–455

Vandekinderen I, Van Camp J, De Meulenaer B et al (2009c) Moderate and high doses of sodium hypochlorite, neutral electrolyzed oxidizing water, peroxyacetic acid, and gaseous chlorine dioxide did not affect the nutritional and sensory quality of fresh-cut iceberg lettuce (*Lactuca sativa* var. *capitata* L.) after washing. J Agric Food Chem 57:4195–4203

Venczel LV, Arrowood M, Hurd M et al (1997) Inactivation of *Cryptosporidium parvum* and *Clostridium perfringens* spores by a mixed-oxidant disinfectant and by free chlorine. Appl Environ Microbiol 63(4):1598–1601

Venkitanarayanan KS, Ezeike OI, Hung Y et al (1999) Efficacy of electrolyzed oxidizing water for inactivating *Escherichia coli* O157:H7, *Salmonella enteritidis*, and *Listeria monocytogenes*. Appl Environ Microbiol 65:4276–4279

Virto R, Manas P, Alvarez I et al (2005) Membrane damage and microbial inactivation by chlorine in the absence and presence of a chlorine-demanding substrate. Appl Environ Microbiol 71:5022–5028

Wang J, Han R (2019) Removal of pesticide on food by electrolyzed water. In: Electrolyzed water in food: fundamentals and applications. Springer, Singapore, pp 39–65

Whangchai K, Saengnil K, Singkamanee C et al (2010) Effect of electrolyzed oxidizing water and continuous ozone exposure on the control of *Penicillium digitatum* on tangerine cv 'Sai Nam Pung' during storage. Crop Prot 29:386–389

White GC (1998) Handbook of chlorination and alternative disinfectants, 4th edn. Wiley, New York, pp 212–287

Wuyun DL (2011) Mechanism and application of slightly acid electrolyzed functional water degradation of organophosphorus pesticide residues. China Agricultural University

Xiong K, Li XT, Guo S et al (2014) The antifungal mechanism of electrolyzed oxidizing water against *Aspergillus flavus*. Food Sci Biotechnol 23(2):661–669

Xuan X, Ling J (2019) Generation of electrolyzed water. In: Electrolyzed water in food: fundamentals and applications. Springer/Zhejiang University Press, Hangzhou

Youssef K, Hussien A (2020) Electrolysed water and salt solutions can reduce green and blue molds while maintain the quality properties of 'Valencia late' oranges. Postharvest Biol Technol 159:111025

Youssef K, Mustafa ZMM, Al-Essawy A (2018) Efficacy of alkaline and acidic electrolysed water generated by some salt solutions against gray mold of table grape: pre and postharvest applications. J Phytopathol Pest Manag 5(1):1–21

Zhang Q, Xiong K, Tatsumi E et al (2012) Elimination of aflatoxin B1 in peanuts by acidic electrolyzed oxidizing water. Food Control 27(1):16–20

Zhang C, Cao W, Hung Y et al (2016a) Application of electrolyzed oxidizing water in production of radish sprouts to reduce natural microbiota. Food Control 67:177–182

Zhang C, Cao W, Hung Y et al (2016b) Disinfection effect of slightly acidic electrolyzed water on celery and cilantro. Food Control 69:147–152

The manufacturer's authorised representative in the EU is Springer Nature Customer Service Centre GmbH, Europaplatz 3, 69115 Heidelberg, Germany. If you have any concerns regarding our products, please contact ProductSafety@springernature.com

Printed and bound by CPI Group (UK) Ltd, Croydon, CR0 4YY
30/09/2026
02225506-0001